SpringerBriefs in Applied Sciences and Technology

Computational Intelligence

Series Editor

Janusz Kacprzyk, Systems Research Institute, Polish Academy of Sciences, Warsaw, Poland

SpringerBriefs in Computational Intelligence are a series of slim high-quality publications encompassing the entire spectrum of Computational Intelligence. Featuring compact volumes of 50 to 125 pages (approximately 20,000-45,000 words), Briefs are shorter than a conventional book but longer than a journal article. Thus Briefs serve as timely, concise tools for students, researchers, and professionals.

Dominique Guérillot

Artificial Intelligence Proxy Models: Applications in Geosciences

Dominique Guérillot
TERRA 3E SAS
Rueil Malmaison, France

ISSN 2191-530X ISSN 2191-5318 (electronic)
SpringerBriefs in Applied Sciences and Technology
ISSN 2625-3704 ISSN 2625-3712 (electronic)
SpringerBriefs in Computational Intelligence
ISBN 978-3-031-90446-2 ISBN 978-3-031-90447-9 (eBook)
https://doi.org/10.1007/978-3-031-90447-9

© The Editor(s) (if applicable) and The Author(s), under exclusive license to Springer Nature Switzerland AG 2025

This work is subject to copyright. All rights are solely and exclusively licensed by the Publisher, whether the whole or part of the material is concerned, specifically the rights of translation, reprinting, reuse of illustrations, recitation, broadcasting, reproduction on microfilms or in any other physical way, and transmission or information storage and retrieval, electronic adaptation, computer software, or by similar or dissimilar methodology now known or hereafter developed.
The use of general descriptive names, registered names, trademarks, service marks, etc. in this publication does not imply, even in the absence of a specific statement, that such names are exempt from the relevant protective laws and regulations and therefore free for general use.
The publisher, the authors and the editors are safe to assume that the advice and information in this book are believed to be true and accurate at the date of publication. Neither the publisher nor the authors or the editors give a warranty, expressed or implied, with respect to the material contained herein or for any errors or omissions that may have been made. The publisher remains neutral with regard to jurisdictional claims in published maps and institutional affiliations.

This Springer imprint is published by the registered company Springer Nature Switzerland AG
The registered company address is: Gewerbestrasse 11, 6330 Cham, Switzerland

If disposing of this product, please recycle the paper.

Introduction

Making new discoveries becomes more and more difficult. Thus several oil companies have decided to focus their efforts on the optimization of the recovery for their mature fields. Different approaches are possible to reach this goal: the improvement of the sweep efficiency; the modification of the oil-in-place viscosity or the injected water for instance (polymer and thermal methods) or the specific well architectures (infill drilling, deviated wells, sidetracks, etc.); the reduction of the remaining oil saturation by chemical processes, etc. All these methods need a better knowledge of the reservoir heterogeneities, knowledge allowing to improve the strategy of oil recovery for existing fields. These heterogeneities are modeled by geostatistical concepts. Recently, innovative solutions, such as the gradual deformation, have been proposed to update these statistical geological models to take into account both the dynamical data (history matching) and the seismic data (reservoir monitoring). But still large uncertainties remain, and experimental design methods are useful to estimate their impact on future productions. Taking into account these uncertainties in the economics of the process is also necessary for a better decision-making process.

The final goal of the methodology described here is to enhance the quality of the production forecasts of hydrocarbon reservoirs. Some history is needed to better understand the motivations of the research work done last 15 years in the area of reservoir characterization, reservoir simulation, and reservoir monitoring. In the '60s, pioneers replaced the 'continuous' reservoir by its discretization by cells to simulate its dynamical behavior using new coming computers. Mechanical equations of transport in porous media (Darcy's equations and continuity equations) were solved numerically through a system of nonlinear equations. In these 'reservoir simulators', the main objectives were to reproduce the physics of the fluid flow in porous media. To simulate the Enhanced Oil Recovery (EOR) processes such as miscible gas injection, thermal methods, polymer injection, Water Alternate Gas (WAG) injection, etc., very sophisticated chemical and physical phenomena were implemented in these reservoir simulators. However, increasing the complexity of these models was not sufficient to be fully confident in the production forecasts. It became increasingly obvious that more geology should be included in these models. In particular, the impact of reservoir heterogeneities and fractures was generally widely underestimated. One of the first breakthroughs was to provide geological models much better described. Three important theories were merged:

1. Geostatistics [1, 2] taking into account the inherent uncertainties in geological models due to the lack of data and indirect measurement of their physical properties.
2. High-Resolution Sequence Stratigraphy integrating the history of sedimentary deposition (mainly variations of the sea level and basin subsidence) [3–5].
3. Structural Geology incorporating regional constraints to inform in the distribution of faults and fractures [6].

The second important breakthrough was to separate this geological description and the petrophysical quantification. This quantification step consists in the allocation to each geological entities (such as lithofacies, faults, fractures, channels, etc.) their petrophysical and acoustic values (porosity, absolute and relative permeabilities of each phase, initial and residual saturations, wettability and capillary pressures, impedance values, velocities, etc.). After this step, a static geological model is described on several millions of cells. This 'continuous' (or as continuous it is possible to represent the geology in computers) aims at representing the best geological view of the reservoir. It should include all available data and geological knowledge. Upscaling these quantified geological models to reach the level of discretization requested for fluid flow calculations is a challenge. A huge amount of work has been done in this area of upscaling non-additive parameters such as permeabilities and capillary pressure curves. The difficulty is that these upscaled values depend on the flow process and are not intrinsic to the medium. This procedure leads to a reservoir simulator grid (100,000 cells to few millions) including, for each cell on the reservoir grid, upscaled parameters. Hence generating grids for heterogeneous reservoirs combined with complex well architectures (deviated wells, side tracks, etc.) is on itself a challenge. Integrated software tools were early proposed to facilitate these three steps: geostatistical modeling, quantification in terms of petrophysical values, upscaling with an adapted reservoir grid [7]. Then come the fluid flow simulation and the production results. In addition to include all the complexity of the physics of Enhanced Oil Recovery Methods, one requires to handle very large amounts of cells with several components.

Parallelization of these calculations is compulsory and recent advances in this area will be briefly described below [8, 9]. These results (mainly, pressure and flow rates at wells) coming after the initial runs are generally quite different from the observed ones. A long and somewhat fastidious work of the Reservoir engineer is to modify this upscaled model to reproduce the observed values. This 'History Matching' process consists in updating the reservoir simulator grid either by a trial-and-error procedure or using gradients methods on the upscaled parameters [10–12]. One of the major innovations described below consists of updating the geological model rather than the upscaled one. This is the only way to enhance the geological model using production data. Updating the upscaled model leads to a reservoir simulator grid not compatible with the initial geological model.

Another important point is to be able to do this updating in a stochastic framework. This methodology passing through will be summed up below. Next generation reservoir simulation platform should include all these previous steps, and an international competition between software providers in Exploration and Production is going on currently. Even after all these steps, uncertainties still remain and a methodology based on experimental design approaches is described.

Quality of the results affects the estimations of recoverable hydrocarbons. However, dealing with such problems cannot only be done using a simulator, as a reservoir simulation can last several hours. To bypass time-consuming reservoir simulations, some authors proposed to use proxies [2, 26]. Several techniques such as polynomial approaches or kriging have been presented in the literature to be used as proxies for reservoir simulators. In most cases, these used proxies are not suited to real problems because they are not adapted to represent nonlinear phenomena. Also, the construction of proxies like kriging or polynomials requires the use of specific experimental designs that affect their quality of prediction. Also, the amount of data needed to design this type of proxy is very high according to the number of input parameters. However, it is possible to construct a neural network with the data already available and to improve the quality of prediction by adding new data.

References

1. G. Matheron, *Éléments pour une théorie des milieux poreux*. Masson, Paris (1967)
2. A.G. Journel, C.J. Huijbregts, Mining Geostatistics. Academic Press, London (1978)
3. G. Matheron, H. Beucher, C. de Fouquet, A. Galli, C. Ravenne, Conditional simulation of the geometry of fluvio-deltaic reservoirs. In: *1st European Conference on the Mathematics of Oil Recovery (ECMOR)*, Cambridge (1987)
4. C. Ravenne, H. Beucher, R. Eschard, P. Houel, Quantification of geological information from outcrops for three-dimensional reservoir modelling. In: *Quantitative Dynamic Stratigraphy*, Springer, Dordrecht (1988). pp. 85–100
5. D. Granjeon, P. Joseph, F. Schneider, Stratigraphic modeling of detrital systems: a forward approach. Bulletin de la Société Géologique de France, **169**(1), 15–25 (1998)
6. M.C. Cacas, E. Ledoux, G. de Marsily, A. Barbreau, P. Calmels, R. Margritta, B. Tillie, M. Duriez, Modeling fractured media by using a stochastic discrete fracture network: Calibration and validation: 1. The flow model. Water Resources Research, **26**(3), 479–489 (1990)
7. D. Guérillot, R. Eschard, C. Ravenne, A. Galli, D. Guellec, B. Doligez, A 3D stochastic modeling system integrating sedimentological and seismic data. In: *11th European Conference on the Mathematics of Oil Recovery (ECMOR)*, London (1989)
8. D. Ricois, R. Eschard, B. Doligez, A. Raveneau, O. Lerat, 3D modelling of complex turbidite reservoirs: the Girassol field (deep offshore Angola). In: *AAPG International Conference*, Barcelona (2003)
9. L. Caillabet, B. Doligez, P. Joseph, Local grid refinement methods for basin modeling–migration modeling. In: *AAPG Hedberg Conference on Basin and Petroleum System Modeling*, El Paso (2004)
10. P. Jacquard, Étude théorique du déplacement d'un fluide miscible dans un autre en milieu poreux. Revue de l'Institut Français du Pétrole, **16**(12), 1781–1801 (1961)

11. P. Jacquard, A.K. Jain, Percolation of miscible fluids within porous media: a theoretical study. SPE Journal of Petroleum Technology, **17**(9), 1025–1032 (1965)
12. J. Antérion, B. Doligez, R. Eschard, Une méthodologie pour une étude intégrée des réservoirs: des données géologiques aux simulations. Pétrole et Techniques,**372**, 43–47 (1989)

Contents

1 **Methodology to Build an Artificial Neural Network for Reservoir Engineering Problems** 1
  1.1 Artificial Neural Network 1
  1.2 Optimal Artificial Neural Network 3
  1.3 How to Generate the Learning Database 4
  References 5

2 **Artificial Neural Networks for Reservoir Engineering Problems** 7
  2.1 Assisted History Matching 7
    2.1.1 Objective Function Involving Artificial Neural Networks 9
    2.1.2 Methodology 10
  2.2 Optimal Well Location Workflow Using an Artificial Neural Network Based Proxy Model 10
    2.2.1 Mean-Standard Deviation Objective Function 14
    2.2.2 Sharpe Ratio 15
    2.2.3 Methodology 15
  2.3 Production Optimization Workflow Using an Artificial Neural Network Based Proxy Model 16
    2.3.1 Mean-Standard Deviation Objective Function 18
    2.3.2 Methodology 19
  2.4 Assessing Uncertainties Using a Proxy Model Base on ANN 20
  References 23

3 **Application to These Advanced Workflows to the Brugge Field Case** 27
  3.1 Description of the Brugge Field 27
  3.2 History Matching 28
  3.3 Well Placement Optimization 30
  3.4 Water Flooding 31
  Reference 32

Thesaurus 33
Annex 1 43
Bibliography 51

Chapter 1
Methodology to Build an Artificial Neural Network for Reservoir Engineering Problems

1.1 Artificial Neural Network

A neural network is based on a simplified model of neuron connected to each other. Each neuron receives inputs and computes an output according to a function defined for each neuron (Fig. 1.1). The function f can be a sigmoid (exponential or tangential), or a Gaussian function, or the sign function. A network of neurons can therefore be represented by the synaptic weights (w) of the different neurons.

The mathematical definition of a neuron with p inputs is a function f defined as following:

$$\left| \begin{array}{l} f : R^{p+1} \times R^p \to R \\ satisfying : \\ 1 : g : R \to R \\ 2 : W \in R^{p+1}, W = (w_1, \ldots, w_p, b) \\ 3 : \forall x \in R^p, f(W, x) = g\left(\sum_{i=1}^{p} w_i z_i + b \right), with\, x = (z_1, \ldots, z_p) \end{array} \right.$$

where, x is the input vector of the neuron, W the weight vector, b_0 is the bias and g is the transfer function.

The collective behavior of a collection of neurons allow to modeling higher order functions in relation to the basic function of the neuron: the multilayer perceptron. The multilayer perceptron is composed of one or more intermediate layers called hidden layers. These hidden layers are composed of several neurons (Fig. 1.2).

A characteristic of neural networks is their ability to learn, such as recognizing a letter, a number, etc. But this knowledge is not acquired a priori. Most neural networks learn with examples. They therefore have an ability to predict, classify and generalize. A neural network will recognize an object more easily because it has often seen it.

© The Author(s), under exclusive license to Springer Nature Switzerland AG 2025
D. Guérillot, *Artificial Intelligence Proxy Models: Applications in Geosciences*,
SpringerBriefs in Computational Intelligence,
https://doi.org/10.1007/978-3-031-90447-9_1

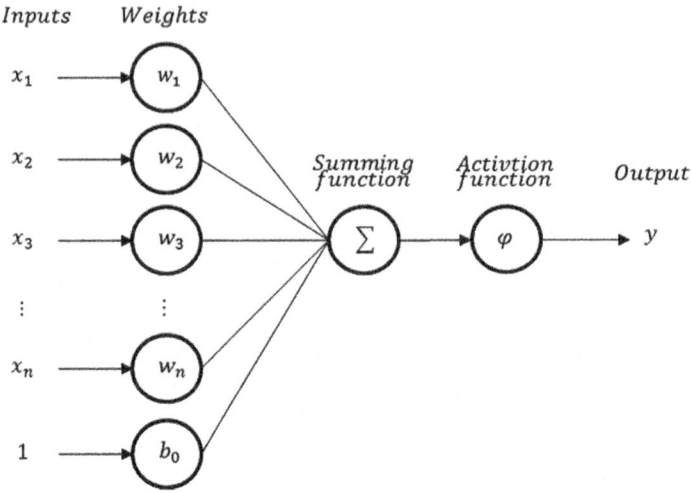

Fig. 1.1 Example of artificial neuron

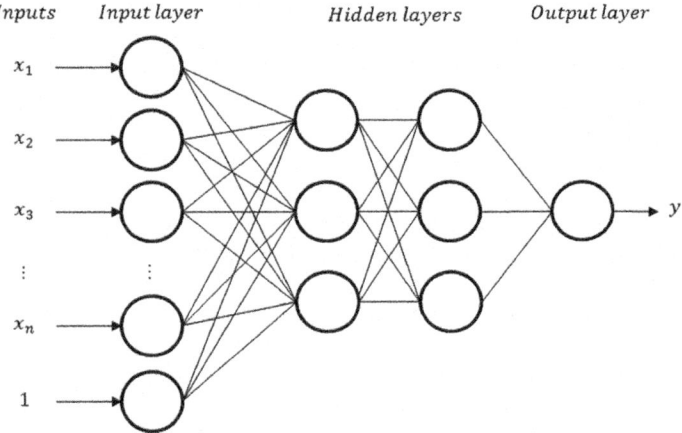

Fig. 1.2 Example of multilayer perceptron with two hidden layers

Neural networks are classified into two classes, supervised learning networks and unsupervised learning networks. For supervised learning networks, inputs and the output that that would be desired for this input are simultaneously presented to the network. For example, a permeability value per facies is presented as input and the value of the accumulated oil at a given date as output. The network must then reconfigure itself, i.e., to calculate its synaptic weights so that the output it corresponds with the most accuracy as possible to the desired output. Back-propagation is a method of calculating weights for a supervised learning network which consists in minimizing the quadratic error of output, which is quite simple when using a differentiable function f (the sigmoid for example), it's why sigmoid functions are used, which is an infinitely differentiable approximation of the Heaviside threshold function. The modification of the weights of the output layer is propagated to the input layer.

A neural network performs a function, which depends on the structure of the network as well as the operation performed by the neurons. Once the learning phase is done, i.e., the weights are stabilized when the examples are presented, the "expert" system can be used as a black box to predict the phenomenon learned when new input parameters are presented.

The universal approximation capacity of neural networks has been demonstrated by several authors [1] and [2, 3] and [4, 5]): "*An arbitrary continuous function, defined on [0,1] can be arbitrary well uniformly approximated by a multilayer feedforward neural network with one hidden layer (that contains only finite number of neurons) using neurons with arbitrary activation functions in the hidden layer and a linear neuron in the output layer*" (Universal approximation theorem; [2]).

Another fundamental property of artificial neural network has been show by Barron (1993): They are "universal parsimonious approximators". "Parsimony" expresses the fact that the network needs a small number of adjustable parameters to mimic any process. For an equal precision, ANN require less adjustable parameters than the other known approximators. For ANN, the number of weights varies linearly according to the number of variables exponentially while for most other variables the number of weights varies exponentially.

Despite these properties, a barrier to the use of neural networks is the determination of the architecture of the ANN, i.e., the number layers and the number of neurons per layer. From a given modeling problem and a learning database, there is no a priori definition of the better architecture. What architecture of ANN gives an optimal generalization?

1.2 Optimal Artificial Neural Network

One of the characteristic of a "good" neural network is not limited to predicting an example used in the learning phase, but a neural network must be predictive for a new case. The evaluation of the model is therefore a paramount step in order to avoid this over-training which is interpreted as a perfect learning of the data.

Among the different approaches available to avoid over-training, the most conventional method is the use of two subsets: a learning set and a validation set. This method of cross-validation [8] allows to stop learning when the error on the validation data begins to increase. This approach allows to a given architecture to stop learning at the appropriate timing.

However, over-learning usually comes from the poor design of the network structure used. To limit the problems of over-learning, it is necessary to determine the adequate number of hidden layers and neurons by hidden layers. To limit the complexity of a neural network we can cite the pruning method which consists in eliminating, once the learning phase is completed, the connections which have the smallest influence on the output error of the neural network [9, 10]. Empirical rules exist to define the number of neurons per hidden layer. The size of the hidden layer must be: equal to that of the input layer; or equal to 75% of this; or equal to the square root of the product of the number of neurons in the input and output layer. A complete review is given by Gnana Sheela and Deepa [11].

The approach used in these briefs to identify the optimal neural network for a given problem is to perform cross-validation on a set of neural networks. This set of neural networks consists of all possible architectures which have a total number of weights less than the number of training data and a number of neurons per hidden layer less than the number of input parameters. The configurations of neural networks contain a maximum of two hidden layers. For each of the configurations, several trainings are carried out starting from different initial synaptic weight values. The chosen architecture will be that of which the normalized root mean squarer error, on the learning and validation data, is minimal.

This approach makes it possible to parallelize the learning phase of each neural network by distributing each network amongst computing nodes.

1.3 How to Generate the Learning Database

To efficiently generate a learning database, a classical approach based on experimental designs (ED) consists of select the data to cover the space of parameters to obtain the most information with a minimum of data [6]. Following the type of proxy model, some experimental designs are more appropriate to fit a model, e.g., Plackett–Burman experimental designs suit to design first-order models, Box-Behnken experimental designs suit to design second-order models. According to the complexity of the model, experimental designs consider or not the interactions between parameters. Here, the Box-Behnken experimental design is used. This type of ED suits to fit quadratic models, considering the interactions between two factors, the linear and squared terms. However, no assumptions about the model to reproduce have been made a priori, and random samples are added to the learning database. The predictive quality of an ANN is dependent on the quality and number of data in the learning base, i.e., the higher the number of learning data, the better the prediction quality of the ANN. The simulations requested to constitute the learning database are independent one to the other and can be quickly done in a parallel manner to reduce the total computing time;

The learning phase aims in to adjust the weights of the ANN in order to the simulated outputs fit the learning outputs. One of the most method used to train an ANN is the back-propagation algorithm [7]. This algorithm consists in to compute the total error between the outputs provided by the ANN and the actual outputs given by the database. The derivative of the error, combined with a learning rate and a momentum term, is used to modify the weights of the ANN. This iterative process is repeated until a minimal error is reached. To avoid the problem of over-learning, i.e.,

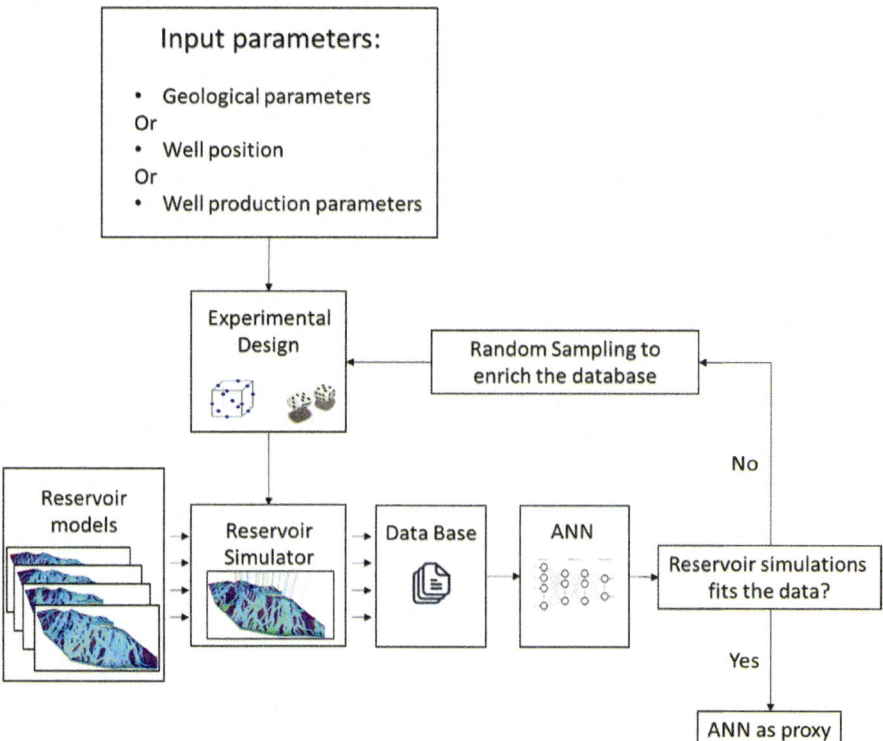

Fig. 1.3 Workflow to generate an artificial neural network

the inability of the network to predict outputs on non-learned data, a validation dataset is used and the learning phase is stopped when the error on the validation dataset increases. As long as the generalization capacity of the ANN is not satisfactory, the learning dataset needs to be improved by adding new samples, e.g., with new random samples, and the learning phase is repeated (Fig. 1.3).

References

1. G. Cybenko, *Continuous Valued Neural Networks With Two Hidden Layers are Sufficient.* Technical Report, Department of Computer Science, Tufts University (1988)
2. G. Cybenko, Approximation by superpositions of sigmoidal functions. Mathemat. Control Signals Syst. (1989)
3. K. Hornik, M. Stinchcombe, H. White, Multilayer feed forward networks are universal approximators. Neural Netw. **2**, 359366 (1989)
4. K. Hornik, M. Stinchcombe, H. White, Universal approximation of an unknown mapping and its derivatives using multilayer feedforward networks. Neural Netw. **3**, 551560 (1990)
5. K. Hornik, Approximation capabilities of multilayer feedforward networks. Neural Netw. **4**, 251–257 (1991)

6. C. Amudo, T. Graf, R.R. Dandekar, J.M. Randle, *The Pains and Gains of Experimental Design and Response Surface Applications in Reservoir Simulation Studies* (Society of Petroleum Engineers, In SPE Reservoir Simulation Symposium, 2009)
7. D.E. Rumelhart, R. Durbin, R. Golden, Y. Chauvin, Backpropagation: the basic theory. in *Backpropagation: Theory, Architectures and Applications* (1995), pp. 1–34
8. C.M. Bishop, *Neural Networks for Pattern Recognition*. Oxford University Press, Oxford (1995)
9. B. Hassibi, D.G. Stork, G. Wolff, Optimal Brain Surgeon: Extensions and performance comparisons. In *Advances in Neural Information Processing Systems*, Morgan Kaufmann, **6**, 88–95 (1993)
10. Y. LeCun, J.S. Denker, S.S. Solla, Optimal Brain Damage. In D. Touretzky (Ed.), *Advances in Neural Information Processing Systems*, Morgan Kaufmann, **2**, 598–605 (1989)
11. K. Gnana Sheela, S.N. Deepa, Review on Methods to Fix Number of Hidden Neurons in Neural Networks. *Mathematical Problems in Engineering* (2013), pp. 1–11. https://doi.org/10.1155/2013/425740

Chapter 2
Artificial Neural Networks for Reservoir Engineering Problems

2.1 Assisted History Matching

History matching of production data is a process requiring a large number of reservoir simulations that are huge time-consuming. To reduce the computation time, an approach consists of using a meta-model to replace the reservoir simulator. Here, a proxy model based on an artificial intelligence technique (artificial neural network) is evaluated and compared to proxy models conventionally used for history matching such as polynomials or kriging methods. The proposed approach provides accurate prediction results to speed up and improve the history matching process. An application to the Brugge field is presented.

In past decades, many methodologies have been proposed to address assisted history matching (AHM) problem (see the reviews of [1] or [2]). Usually, petroleum reservoir modeling is done with geostatistical methods to represent the spatial distribution of reservoir properties [3, 4]. The uncertainties related to the spatial distribution of reservoir properties are taken into account by the use of a set of realization of the geostatistical model. The history matching process aims to update the set of realization to fit the available data. The first AHM considered only production data [5]. Then well data has been combined with seismic data, this problem is called seismic history matching (for example, [6, 7]). The number of parameters being upper than the number of available data, history matching is an ill-posed problem, many solutions can lead to the same fitness of data history. Parametrization approaches have been proposed to reduce the number of parameters [8, 9]. Whatever the formulation of the history matching problem, the problem is solved with an optimization method. The algorithm used to solve this inverse problem can be split into two categories: gradient-based method and gradient-free methods. Gradient-based methods are popular due to their fast convergence properties [10] when the initial guess is close to a minimum. However, following the initial guess, the optimization process may lead to local minima which not as optimal that would be the global minimum. To reach a better solution avoiding these local minima, many authors proposed to apply free-gradient-free algorithm. However, such methods have known to be much time-consuming in term of the number of resolutions of direct problem, i.e., the number of reservoir simulations. To work around the matter of the number of reservoir simulator, an approach consists in using surface response models which aims to reproduce the behavior of the reservoir simulator but in a 'cheaper' way.

Surface response models, also called metamodels, surrogate model or proxy model, are prevalent to speed up the optimization processes because these approaches allow approximating the exact models but at a much lower CPU cost. This CPU cost becoming negligible, many forward runs Many applications have already published for applications in reservoir simulation [11], for example in history matching problems [12–14], production optimization [15], uncertainty analysis [16] and well placement problems [17–20].

Among the most used proxy models, we can mention polynomials, kriging, or artificial neural networks. Also, artificial neural networks have shown mainly their superiority compared to conventional methods. In fact, ANNs are mathematical models able to approach, with the desired precision, any continuous function [21-25] and it is also what allows them to be applied in a very efficient way for many models in fields as varied as the aerospace for the automatic control, the defense for the facial recognition.

We focus here on an artificial neural network, an Artificial Intelligent (AI) technic. Since many years and the continued increase of computing power, AI technics has become popular in reservoir engineering. Guérillot [26] presented an approach based on a fuzzy logic concept to identify the Enhanced Oil Recovery (EOR) method to apply following the reservoir characteristics. In many history matching and optimization studies, the reservoir simulator has been substituted by an artificial neural network to speed up the processes [12–15, 27–29]. Other authors proposed to replace only a part of the reservoir simulator, e.g., the thermodynamic equilibrium calculation being the most consuming time part in reservoir simulation Guérillot [30] replaced only the thermodynamic solver with an ANN. Guérillot [31] replaced the geochemical solver in reactive transport simulator. In the same way, Wang et al. [32] use an ANN to accelerate the vapor–liquid equilibrium.

Here, an ensemble of geological realization is considered. The proxy model used is an optimal artificial neural network (ANN). An ANN is designed for each reservoir simulator output for each well. The outputs of the proxies are the average and standard deviation value of simulated data of the ensemble of the geological realizations. The objective function to minimize is a combination of these proxies. The optimization is done using a global optimization method coupled with the objective function composed of the ANN outputs. When the optimization process has converged, the solution provided by the optimizer with the proxy models is compared with the results obtained with the reservoir simulator. If the solutions are different, proxy models must be improved with additional training data points. Then, new proxy models are designed, and optimization is executed.

First, the formulation of the objective function considering the uncertainties of the geological model is recalled. Then, the methodology to address the history matching problem using ANN as proxy models is detailed. An application of this methodology is presented with the Brugge field.

2.1.1 Objective Function Involving Artificial Neural Networks

The general formulation of the objective function to address history matching problems consists of minimization between observed data and the simulated data. Different formulations can be applied, e.g., root mean squared error (RMSE), normalized RMSE, maximum error. A review of the different objective function formulation is given by Mata-Lima [33].

Here, we consider an ensemble of geological models, generated following geostatistical laws, supposed to represent the uncertainty related to the distribution of petrophysical parameters. The objective function is defined as the root mean of the average squared error between observed data and simulated data for the set of geological realization.

$$J(X) = \sqrt{\frac{1}{N.P} \sum_{t=1}^{N} \sum_{i=1}^{P} \mathbb{E}_\Omega \left[\left(d_i^t(X) - d_{obs,i}^t \right)^2 \right]} \quad (2.1)$$

where X is a vector of optimization variables (e.g., grid block porosity, permeability), N is the number of observed data, P is the number of well (producers and injectors), $\mathbb{E}_\Omega$ is the average operator, d_i^t is a simulated data, $d_{obs,i}^t$ is an observed data (e.g., bottom hole pressure, oil production rate, water production rate).

Following the properties of the variance, (2.1) can be rewritten as follow:

$$J(X) = \sqrt{\frac{1}{N.P} \sum_{t=1}^{N} \sum_{i=1}^{P} \sigma_\Omega^2 \left[d_i^t(X) \right] + \left(\mathbb{E}_\Omega \left[d_i^t(X) \right] - d_{obs,i}^t \right)^2}, \quad (2.2)$$

where σ_Ω is the standard deviation operator.

For a deterministic model, the formulation (2.2) of the objective function is equaled to the classical RMSE.

The methodology aims to address the history matching problem considering the uncertainties related to the spatial distribution of geological parameters. The classical approaches imply to calculate the average error on production data for an ensemble of geological realization at each optimization step until convergence of the algorithm. For example, if the geological uncertainties are taken into account through the consideration of 100 reservoir model realizations, at each time step of the optimization 100 reservoir simulations must be done to calculate the error between production data and the average value given by the 100 reservoir simulation. This type of approach is very expensive in calculation time.

Here, the average and standard deviation resulting from an ensemble of reservoir simulations will be replaced by an ANN. For each time step and each data, a proxy model will provide an approximation of these terms.

2.1.2 Methodology

The presented methodology consists in to build an artificial neural network for each reservoir simulator output, i.e., for each type of data at each time step. The proxy models provide the average and standard deviation of each production data. The procedure to build the proxy models is: (1) one-first set of simulation is generated using experimental designs, and random sampling; (2) proxy models of each reservoir simulator output for each well are designed using the first set of simulations; (3) the evaluation of each proxy is done with a new set of random sampling. If the accuracy of the proxies is satisfactory, i.e., the error between the outputs given by the ANNs and the outputs given by the simulator are small, these proxies are used to define the objective function used to solve the history-matching problem.

When the ANN learning is finished, i.e., the ANN has acceptable generalizability, these outputs are used to define an approximation of the objective function to be minimized. The objective function then becomes:

$$\tilde{J}(X) = \sqrt{\frac{1}{N.P} \sum_{t=1}^{N} \sum_{i=1}^{P} \left(ANN_{\sigma\Omega}^{d_i^t}(X)\right)^2 + \left(ANN_{\mathbb{E}\Omega}^{d_i^t}(X) - d_{obs,i}^t\right)^2}, \qquad (2.3)$$

where $ANN_{\mathbb{E}\Omega}^{d_i^t}(X)$ is the first output of the ANN corresponding to the average of an output data for a given input X and $ANN_{\sigma\Omega}^{d_i^t}(X)$ is the second output of the ANN corresponding to the standard deviation of an output data for a given input X;

The optimization problem to solve consists in determining the value X^* which minimizes the objective function (Fig. 2.1):

$$\tilde{J}(X^*) = \min_{X \in U} \tilde{J}(X). \qquad (2.4)$$

2.2 Optimal Well Location Workflow Using an Artificial Neural Network Based Proxy Model

Planning the position of a new well is an essential task to increase hydrocarbon production. However, trying to capture the uncertainty related to the geological properties, an ensemble of geological realizations of the models must be considered. The objective can be to identify the locations of the new wells that maximize

2.2 Optimal Well Location Workflow Using an Artificial Neural Network ...

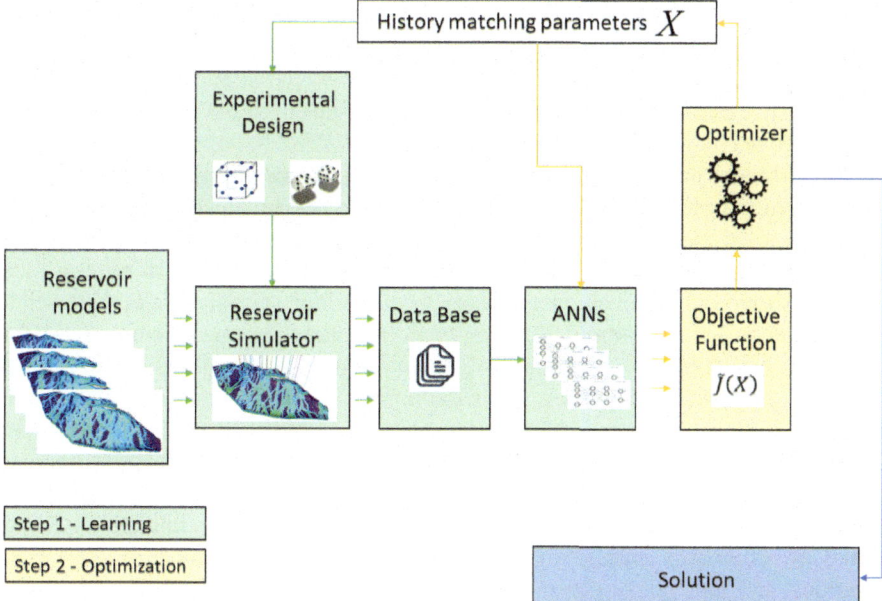

Fig. 2.1 Flowchart of the history matching methodology: (1) Learning phase to design the artificial neural networks; (2) Minimization of the approximation of the objective function using the ANNs; (3) The solution is checked using the reservoir simulator

the average oil recovery or Net Present Value (NPV), for an ensemble of geological models. Also, considering an ensemble of geological models, an objective can be to minimize the standard deviation of the NPV (or cumulative oil recovery). Reducing the uncertainty on the NPV leads to reduce the risk for the placement of the new wells. These two objectives can be defined with a "mean-standard deviation" formulation of the objective function. Following the weight given to each objective, the placement of the wells is not the same. To know all the well-placement possibilities and associated risk, many optimization problems must be done. Such as procedures, require the use of optimization method and reservoirs simulations on each reservoir model. This optimization problem leads to a high number of reservoir simulation to determine the appropriate location and can be expensive in computation time. To reduce the computation time, a method consists in to substitute the reservoir simulator by a meta-model. Here, the use of an artificial neural network as a proxy model is considered. The objective is to propose a method to speed up the well-placement optimization process using a proxy model based on an artificial intelligence technique to replace the reservoir simulator. The proposed methodology considers an ensemble of realization and aims to provide an overview of solutions in an economic point of view inspired by portfolio optimization problems. An application of the methodology is presented for the Brugge field.

The localization of new wells is a challenging task in the oil field development plan. Many authors are proposing to maximize the oil recovery or the Net Present Value (NPV) by controlling the location of the new wells. Some of them consider a deterministic reservoir model while others consider an ensemble of reservoir models to take into account the geological uncertainty.

This problem has been addressed considering the reservoir model as deterministic. Gradient-based methods have been applied to maximize NPV [34, 35]. From a mathematical point of view, well-placement optimization is a challenging task due to the complexity of the objective function usually composed of several local minima. To avoid the convergence to these local minima, several authors proposed to use derivative-free algorithms. Emerick et al. [66] had developed a computer-aided optimization tool based on a genetic algorithm to optimize well-placement considering arbitrary well trajectories. Bouzarkouna et al. [36] proposed to couple the Covariance Matrix Adaptation Evolution Strategy (CMA-ES) [37] optimizer with surface response models of the NPV for each well. This approach leads to improve the convergence of the optimization method knows to be costly in term of direct evaluation of the objective function, i.e., the number of reservoir simulations. Isebor et al. [38] proposed a method to optimize simultaneously the number of wells, their locations, and the control parameters using a derivative-free algorithm. Awotunde and Sibaweihi [39] used the CMA-ES and Differential Evolution (DE) algorithms and considered a multiobjective approach composed of the NPV and the Voidage-Replacement Ration (VRR). Forouzanfar et al. [40] used the CMA-ES to optimize the well-placement and the control parameters sequentially or simultaneously for a deterministic geological model. Alrashdi and Sayyafzadeh [41] compared $(\mu + \lambda)$ Evolution Strategy, Genetic Algorithm (GA), Particle Swarm Optimisation (PSO) and CMA-ES for well control, placement, and joint optimization problems.

To compensate for insufficient information on the structure of the reservoirs at distances less than the spacing of the soundings, an approach proposed by geologists consisted in reasoning by analogy [3, 42–45]: to look out for geological structures comparable to those of the oil formations that interest us. The fundamental hypothesis that is made in this step can be expressed as follows: The way the sediments have been deposited can be explained, and these depositing mechanisms have been repeated for different deposits: there are families of reservoirs that have been built "roughly" in the same way. Thus, if we can reconstruct a well-known deposit, if we can characterize it with only a few parameters, then we can reconstruct other deposits of the same type of deposit, using available information (electrical logs, well tests, s surveys, etc.). The probabilistic character of these "images" of reservoirs is the ransom of the uncertainty that exists in an immense region where little is known about the information available in the wells. For heterogeneous deposits, one possible approach is to assume different geometries for a deposit and to make each one of them reservoir studies. Of course, the number of these assumptions can only be limited in number, given the time it would take to interpret the calculations, which themselves are very heavy. But the remaining deposits to be exploited are becoming more and more difficult and fewer and fewer, and the costs of calculations continue to fall considerably. Computers are getting

2.2 Optimal Well Location Workflow Using an Artificial Neural Network ...

faster and faster and allow an increasing amount of data to be processed. Thus, three-dimensional representations of the reservoirs are now simulated by geostatistical methods integrating data from various sources outcrop, wells, seismic, dynamical data.

In the well-placement optimization context, Güyagüler and Horne [46] used the utility framework with Genetic Algorithm (GA) to maximize the NPV. Wang et al. [47] proposed to maximize the NPV using a Retrospective Optimization (RO) procedure consisting in increasing the number of geological realization sequentially at each optimization iteration.

Within the well-placement optimization process, most the works are focused on the maximization of the NPV and consider the risk only as an input variable due to geological uncertainties without thinking of the risk in an economic point of view. Here, we take place in the concept introduced by the portfolio optimization problems in finance that consists of analyzing a 'mean-standard deviation' model of an objective function [48]. Couet et al. [49] apply this methodology to optimize the water-flooding strategy. Capolei et al. [50] propose a mean–variance objective function and evaluate the 'risk-expected return' through the Sharpe index [51].

Whatever the formulation of the objective function, the well-placement optimization is a process involving a large number of reservoir simulation, moreover if geological uncertainties are taken into account through many realizations of the geological model. In this context, a classical approach consists in using of response surface model to bypass the reservoir simulator. Proxy models (e.g., kriging, polynomial, and artificial neural network) have been widely used for reservoir engineering problems such as history-matching, production forecasting, and production optimization. With the continued increase in computer power, artificial intelligence techniques have rapidly met interest in the reservoir engineering community. In 1988, Guérillot proposed an Expert System (ES) based on the concept of fuzzy logic to identify the most suitable Enhanced Oil Recovery (EOR) process to apply in a given reservoir. Artificial neural networks (ANN) to replace the reservoirs simulator has been applied to optimization processes [12–15, 27–29]. Guérillot [30] proposed a workflow to replace the calculation of thermodynamic equilibrium in reservoir simulation by an ANN. Guérillot and Bruyelle [31] proposed to speed-up reservoir simulations with an ANN as a proxy model of the geochemical equilibrium solver in a reactive transport simulator.

In this work, the methodology proposed by Bruyelle and Guérillot [18–20] for water flooding optimization is applied for well-placement optimization. This approach is in line with the concept introduced by the portfolio optimization problems in finance that consists of analyzing a 'mean-standard deviation' model of the NPV. This approach requires a considerable number of reservoir simulation to optimize the production strategy for different formulations of the objective function. An artificial neural network has been used to provide the average and standard deviation of the NPV and also to bypass the use of the reservoir simulator. The proposed approach allows evaluating of all the formulation of the objective function to give the optimal production strategies with the "risk-expected return" index associated. The risk is considered by the standard deviation of the NPV for an ensemble of geological realizations.

Here, first, the "mean-standard deviation" objective function formulation for the well-placement optimization problem and the Sharpe ratio are recalled. Then, the methodology that consists in to replace the NPV calculations (mean and standard deviation) by an artificial neural network is presented. The methodology to design the artificial neural network predicting the average and standard deviation of the NPV is described. Finally, an application of this method for well-placement optimization to the Brugge field is proposed.

2.2.1 Mean-Standard Deviation Objective Function

A classical formulation of the Net Present Value (NPV) in a well-placement optimization context is defined as follows:

$$NPV(X, T) = \sum_{i=1}^{T} \Delta_{t_i} \frac{Q_o r_o - Q_w r_w - Q_I r_I}{(1+d)^{t_i}} - \sum_{j=1}^{N} C_j, \quad (2.5)$$

where X is a vector of optimization variables (i.e., well location, well production/injection parameters), T is the number of time steps, Δ_{t_i} is a time step, Q_O is the oil production rate, Q_W is the water production rate, Q_O is the water injection rate, r_O is the oil price, r_w is the water production cost, r_I is the water injection cost, d is an annual discount rate, t_i is the cumulative time, N is the number of new well and C_j is the drilling cost applied to new wells.

Uncertainty is considered through the use of a set of reservoir models representing the variations of the geological properties. For all these models, we suppose that the calibration of the history of production data has been done. The objective is to provide an optimal localization of new wells in term of return/risk ratio. The multi-criterion problem to solve consist of maximizing the average of the NPV and minimizing the standard deviation of NPV. The aim is to find $X^* \in U$ such as:

$$J(X^*) = \max_{X \in U}(\alpha \mathbb{E}_\Omega[NPV(X)] - (1-\alpha)\sigma_\Omega[NPV(X)]), \quad (2.6)$$

where X is a vector defining the optimization variables (i.e., well location, well production/injection parameters), U is the search space, α is a coefficient between [0, 1], Ω is a set of geological realizations, $\mathbb{E}_\Omega$ is the average operator and σ_Ω is the standard deviation operator.

Following the value of α, the objective function advantage the maximization of the NPV or the minimization of the standard deviation. A value of α equals to one correspond to the classical formulation of the objective function, i.e., maximization of NPV. Increase the value of α allows reducing the standard deviation of NPV and decreasing the risk.

2.2.2 Sharpe Ratio

Sharpe [51] defined the performance of one solution respect to its risk with an indicator measuring the ratio of the gain expected and the standard deviation of the gain.

$$S_h(X) = \frac{\mathbb{E}_\Omega[NPV(X)] - r_f}{\sigma_\Omega[NPV(X)]}, \qquad (2.7)$$

where r_f is a benchmark return of a risk-free configuration.

Here, the benchmark return of a risk-free configuration is assumed to be the one obtained for the objective function with, i.e., the configuration that minimizes the standard deviation. The Sharpe ratio will be computed only for values of α greater than zero.

This ratio is an indicator of the performance of a given solution and can be interpreted as follow. A negative value designates a solution with poorer performance than the least risky configuration, i.e., the configuration that minimizes the standard deviation and corresponds to the solution obtains for the objective function with $\alpha = 0$. A ratio value between 0 and 1 designates a solution with a risk too high compared to the benefit expected. A ratio upper than 1 designates a solution with a high expected benefit for a low taken risk.

In an economic context, the best solution should be the one that maximizes the Sharpe ratio. Thus, the definition of α does not matter. However, in a real industrial application, the operators want to know a maximum of information on the gain and risk taken. The Sharpe ratio must be calculated for different value of α, i.e., the well-placement optimization must be done for each value of α.

2.2.3 Methodology

The proposed methodology aims to provide an overview of the possible solutions with the associated "risk-expected return" associated. This workflow implies to perform an optimization process for each formulation of the objective function, i.e., for different value of α. Given the geological uncertainty resulting from a set of geological realizations, each optimization process is all the longer as the number of reservoir models is significant. To reduce the computation time, an artificial neural network as a proxy model is used. The use of a proxy implies to generate a database representing the problem to reproduce. The ANN is then trained using the learning database. An approximation of the objective function is designed with the outputs of the ANN. Global optimizations are done for each value of α. To finish, an overview of the solutions with the associated Sharpe index is done. The steps of the methodology are described below.

The objective function is defined with the outputs of the ANN, i.e., the average and standard deviation of the NPV. The problem to solve is to identify $X^* \in U$ such as:

$$\tilde{J}(\alpha, X^*) = \max_{X \in U}\bigl(\alpha ANN_{\mathbb{E}\Omega}(X) - (1-\alpha)ANN_{\sigma_\Omega}(X)\bigr), \qquad (2.8)$$

where $ANN_{\mathbb{E}\Omega}(X)$ is the first output of the ANN corresponding to the average of the NPV for a given input X and $ANN_{\sigma_\Omega}(X)$ is the second output of the ANN corresponding to the standard deviation of the NPV for a given input X;

For each value of α, a well locations solution is given with the associated average and standard deviation of the NPV. The ensemble of the couples "average—standard deviation" of the NPV form the efficient frontier, i.e., the location of the wells that maximizes the average NPV for a given level of risk. For each solution, the Sharpe ratio is given.

2.3 Production Optimization Workflow Using an Artificial Neural Network Based Proxy Model

Water flooding is the main technic to recover hydrocarbons in reservoirs. For a given set of wells (injectors and producers), the choice of injection/production parameters such as pressures, flow rates, and locations of these boundary conditions have a significant impact on the operating life of the wells. As a large number of combinations of these parameters are possible, one of the critical decision to make is to identify an optimal set of these parameters. Using the reservoir simulator directly to evaluate the impact of these sets being unrealistic considering the required number of simulations, a common approach consists of using response surfaces to approximate the reservoir simulator outputs. Several techniques involving proxies model (e.g., kriging, polynomial, and artificial neural network) have been suggested to replace the reservoir simulations. These briefs focalize on the application of artificial neural networks (ANN) as it is commonly admitted that the ANNs are the most efficient one due to their universal approximation capacity, i.e., capacity to reproduce any continuous function. These briefs presents a complete workflow to optimize well parameters under water flooding using an artificial neural network as a proxy model. The proposed methodology allows evaluating different production configurations that maximize the NPV according to a given risk. The optimized solutions can be analyzed with the efficient frontier plot and the Sharpe ratios. An application of the workflow to the Brugge field is presented in order to optimize the water flooding strategy.

Many papers are proposing to optimize the water flooding processes when injection water to recover the hydrocarbon in place. Most of them aim to maximize the oil recovery or the Net Present Value (NPV) by controlling the pressure/rate of the wells or smart wells equipped with Inflow Control Valves (ICV). This type of approach has been successfully applied to deterministic models [52–54]. Bittencourt and Horne [55] proposed a genetic algorithm for the development reservoir. Asadollahi and Naevdal [56] utilized a gradient-based method to maximize the NPV and compare results obtained following three different control parameters (oil production rate, liquid production rate, and bottom hole pressure). The method has been applied on an average model resulting from a history-matching process with the Ensemble Kalman Filter (EnKF) procedure, and the solution of the optimization has been validated on random history-matched models. Pinto et al. [57] optimized the ICV parameters to maximize the NPV with genetic algorithms. Forouzanfar and Reynolds [67] proposed a methodology using a gradient-based algorithm to solve the general well placement optimization problem, i.e., optimization of the number of wells, locations, and production rates.

Usually, in static reservoir modeling, the uncertainty assessment related to the spatial distribution of the petrophysical properties is done through geostatistical models [3, 4]. Van Essen et al. [58] applied a gradient-based method to optimize the average NPV using an ensemble of realization of the geological model. Alhuthali et al. [59] proposed a methodology based on the minimization of the difference of arrival time of the waterfront in all producer wells using streamline models and multiple geological realizations. The method has been applied to wells equipped with ICV. Such an approach makes it possible to sweep the field efficiently and thus maximize hydrocarbon production. Chen et al. [60] proposed an Ensemble-Based Optimization (EnOpt) procedure combined with the EnKF to maximize the NPV. Yasari et al. [61] proposed a multiobjective formulation solved with a derivative-free optimization method.

Couet et al. [49] considered the financial and reservoir uncertainty with a mean-standard deviation objective function inspired by the portfolio optimization problems in finance [48]. In the same way, Capolei et al. [50] proposed a mean–variance objective function and used the Sharpe ratio to evaluate the risk-expected return. In these approaches, the risk is examined through the standard deviation or variance of the NPV for an ensemble of geological realizations. Hørsholt et al. [62] presented an implementation of the mean–variance formulation using industrial tools (Eclipse 300 and Matlab) and a water flooding application on a synthetic 2-dimensional black-oil reservoir model with 30 realizations representing the uncertainty associated to the permeability field. Fu and Wen [63] used the inverse of the Sharpe ratio as an indicator of the risk for production optimization problems.

This briefs is in line with the approach considering the risk due to the standard deviation of the NPV. The performance of the optimized production strategies concerning its risk is evaluated with the ratio defined by Sharpe [51]. Compared to previous works [49, 50], a mean-standard deviation formulation of the objective function combined with an artificial neural network (ANN) as a proxy model of the NPV is proposed to speed up the optimization process. The optimization process is done with the Covariance Matrix Adaptation Evolution Strategy (CMA-ES) algorithm [37]. This derivative-free method is a global optimization method that allows avoiding local minima of the objective function and reaches a better solution than the gradient-based techniques. The CMA-ES method has been applied for History-Matching (Bruyelle and Lange 2009) and well placement optimization [64].

Since many years, Artificial Intelligence (AI) technics have been used in the oil and gas industry. For Enhanced Oil Recovery (EOR), Guérillot [26] proposed to use an Expert System (ES) based on the fuzzy logic concept to assist in the selection of the EOR process to apply. Artificial neural networks have been used for reservoir engineering problems to replace the reservoirs simulator. For optimization problems, Johnson and Rogers [65] combined artificial neural networks with a genetic algorithm to optimize the placement of wells. Cullick et al. [12] proposed to use an ANN to replace the fluid flow simulator in order to find the values of the attributes to be used as an initial estimation point for the optimization loop with the actual simulator. Silva et al. [27] had compared different architectures of ANN for history matching problems. Costa et al. [13] used an ANN as a proxy to reproduce the objective function in an optimization process and show that ANN captures the nonlinearities of the problems. Foroud et al. [14] presented an application of ANN combined with a genetic algorithm as an optimization method for an assisted history matching study. Bruyelle and Guérillot [28] proposed to compute the derivative of ANN to optimize an objective function using a gradient-based method. Guérillot [30] proposed a workflow to replace the calculation of thermodynamic equilibrium in reservoir simulation by an ANN. Guérillot and Bruyelle [29] and [15] proposed a complete methodology addressed to history matching and production forecast problems using an optimal artificial neural network as a proxy model coupled with a global optimization method. Guérillot and Bruyelle [31] replaced the geochemical equilibrium solver by an ANN to speed-up the reactive transport simulations.

After the description of the mean-standard deviation objective function for the production optimization problem, the Sharpe ratio used to evaluate the performance of the optimized production strategies concerning its risk is recalled. Then, the methodology, including the use of an artificial neural network as a proxy model of the NPV function, is described. The methodology consists of design an artificial neural network enables to predict the average and standard deviation of NPV for an ensemble of geological realization representing the uncertainties. Global optimizations are performed on a surface response model composed of the outputs of the ANN and a parameter allowing to give more or less importance to the risk, i.e., the standard deviation of the NPV. Finally, an application of this approach for water flooding optimization to the Brugge field is presented.

2.3.1 Mean-Standard Deviation Objective Function

The classical formulation of the Net Present Value (NPV) in a production optimization context is defined as follows:

$$NPV(X, T) = \sum_{i=1}^{T} \Delta_{t_i} \frac{Q_o r_o - Q_w r_w - Q_I r_I}{(1+d)^{t_i}}, \quad (2.9)$$

where X is a vector of control parameters (e.g., well production/injection parameters), T is the number of time steps, Δ_{t_i} is a time step, Q_O is the oil production rate, Q_W is the water production rate, Q_O is the water injection rate, r_O is the oil price, r_w is the water production cost, r_I is the water injection cost, d is an annual discount rate and t_i is the cumulative time.

We take place in the framework of an optimization of the production strategy under uncertainty. Uncertainty is represented by a set of geological models for which the calibration of the history of production data has been made. The objective is to propose an optimal production strategy in term of return/risk ratio. The problem to solve can be defined as a multi-criterion objective function where the average of NPV is maximized, and the standard deviation of NPV is minimized. The optimization problem consists of identifying $X^* \in U$ such as:

$$J(X^*) = \max_{X \in U}(\alpha \mathbb{E}_\Omega[NPV(X)] - (1-\alpha)\sigma_\Omega[NPV(X)]), \qquad (2.10)$$

where X is a vector defining the well production/injection parameters, U is the search space, α is a coefficient between $[0, 1]$, Ω is a set of realizations, $\mathbb{E}_\Omega$ is the average operator and σ_Ω is the standard deviation operator.

A value of alpha equal to one leads to maximize the NPV without worrying about risk due to uncertainty, i.e., the risk of reaching that profit is maximized. On the other hand, a value of alpha equal to zero leads to minimizing the standard deviation of the NPV, i.e., the NPV is not maximized, but the risk of not reaching this NPV value is reduced. Such a formulation of the objective function (2.2) leads to the question of the choice of the value of the alpha parameter.

2.3.2 Methodology

The optimization process requires performing a flow simulation on all geological models to evaluate each production configuration. The higher is the number of production control parameters, the higher is the number of combinations of the control parameters. Also, according to the chosen optimization method (gradient-based method, genetic algorithm, etc.), the number of simulations to converge towards a solution can become too important. In this case, the engineer is often constrained by the computation time, and he could not evaluate the objective function for different levels of risk, i.e., different values of alpha.

The outputs of the ANN are used to compute the objective function. The formulation of the problem to solve consist of identifying $X^* \in U$ such as:

$$\tilde{J}(\alpha, X^*) = \max_{X \in U}\bigl(\alpha ANN_{\mathbb{E}_\Omega}(X) - (1-\alpha)ANN_{\sigma_\Omega}(X)\bigr), \qquad (2.11)$$

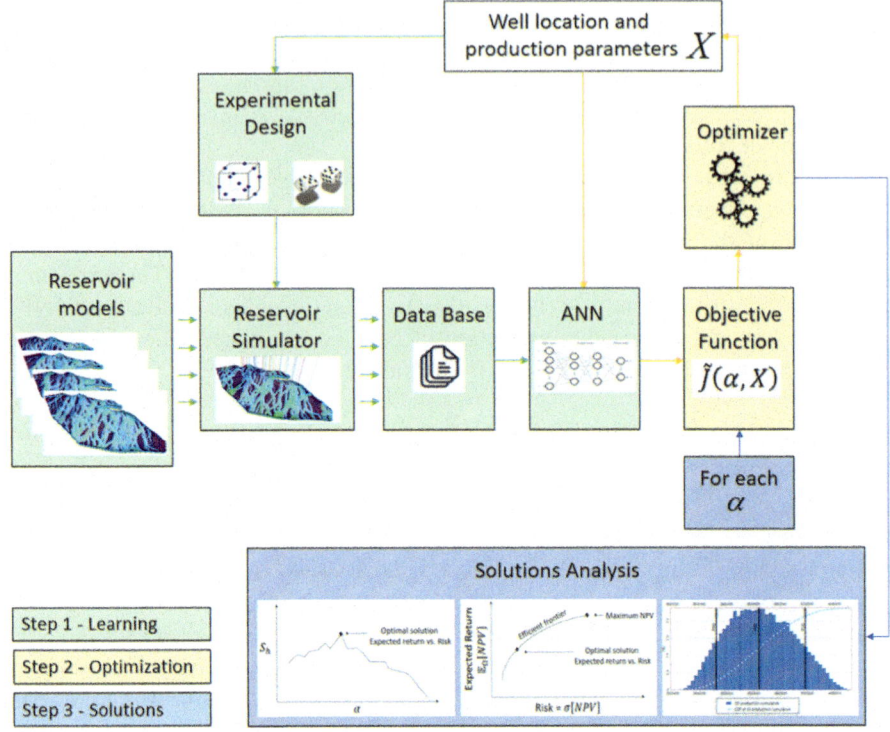

Fig. 2.2 Flowchart of the well placement methodology: (1) Learning phase to design an artificial neural network; (2) Optimization of well-placement for different value of α using the ANN; (3) Solution analysis to identify the best "risk-expected return" solution

where $ANN_{\mathbb{E}\Omega}(X)$ is the first output of the ANN corresponding to the average of the NPV for a given input X and $ANN_{\sigma_\Omega}(X)$ is the second output of the ANN corresponding to the standard deviation of the NPV for a given input X;

Figure 2.2 presents the flowchart of the methodology. Step 1 that corresponds to the generation of the learning database can be parallelized by distributing the reservoir simulations on computing nodes. The optimization is done on a response surface model. The solutions are obtained instantaneously, that allows evaluating the objective function for many values of α (Fig. 2.3).

2.4 Assessing Uncertainties Using a Proxy Model Base on ANN

Decisions for field development of oil and gas reservoirs are often based on uncertainties assessment on forecast productions and other variables which are highly impacted by the uncertainties on the reservoir characteristics. Using geostatistical models, it

2.4 Assessing Uncertainties Using a Proxy Model Base on ANN

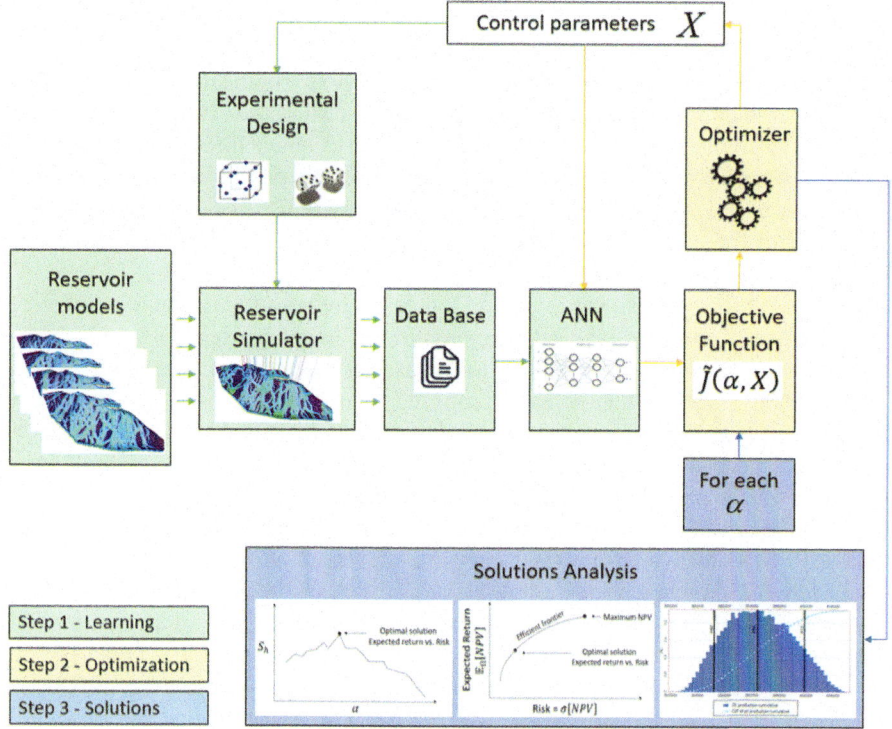

Fig. 2.3 Flowchart of the methodology: (1) Learning phase to design an artificial neural network; (2) Optimization of production for different value of α using the ANN; (3) Solution analysis to identify the best "risk-expected return" solution

would require thousands of flow simulations of several hours each to consider the geological uncertainties. Each of these simulations would require several hours even with current high power computers. To bypass this restriction due to the computation time, one approach consists to replace the simulator by an approximation of it, also called proxy. These briefs focus on the use of Artificial Neural Networks (ANN) proposing an innovative method to build an optimal ANN.

Many tools are used to model geological reservoirs in order to perform simulations that aim to predict the reservoir behavior subject to production constraints (position of wells, injection/output rate, etc.). Also, these models are designed from numerous and various type of data from a wide range of scales (seismic data, well data, outcrop observations, etc.). These data are limited and subject to errors. Integrating these data to build a model does not accurately characterize all reservoir parameters. The model generated is the result of a series of assumptions made on the parameters and their distributions, for example to represent the reservoir heterogeneities.

The history matching phase, which consists in constraining the model to the production data (pressure, flow rate, etc.), allows to reduce the range of uncertainties of several parameters. Generally, the history matching step leads to identify several distinct models but these models are concordant on their capacities to reproduce the production history of the reservoir.

These briefs focus on the next stage of the reservoir engineering workflow after the history matching phase, i.e., the production forecast phase. The different models generated from the history matching phase generally behave differently, resulting in distinct prediction forecasts and this in spite of the reduction of uncertainties on the parameters of the model. During the reservoir production, a crucial step is to assess the impact of the uncertainties arising from the parameters of the geological and/or reservoir model on production forecast.

Uncertainty assessment in production forecast requires a so large number of reservoir simulation runs that it is often impossible to use the reservoir simulator due to the computation time. Therefore, to bypass time consuming reservoir simulations, several techniques have been proposed to mimic the reservoir simulations using proxies, also known as Response Surface Models (RSM). The first proxies which were proposed were built using linear and bilinear polynomials of the input parameters (Damsleth et al. 1991, Zabalza-Mezghani et al. 2004). It appears that polynomials gave 'smooth' outputs and were not always adapted to the non-linear behavior of the results of the reservoir simulator such as water cuts for instance. For improving these proxies, various enhancement has been proposed. Slotte and Smorgrav (2008) propose a proxy model based on polynomials combined with multi-dimensional kriging to reproduce the output of a fluid flow simulator. Other proxies are described in Babaei and Pan [11] who present a review of the use of RSM in petroleum industry. Babaei and Pan (2016) had compared several proxy models to optimize water injection (cubic radial basis functions, kriging and multivariate adaptive regression splines).

After using ANN for history matching [29], this text focuses on the use of ANN as proxy model for the uncertainty assessment in production forecast. ANN have been also used for reservoir engineering problems to mimic the reservoirs simulator. Silva et al. [27] had compared different architectures of ANN for history matching problems. Costa et al. [13] used ANN as proxy to reproduce the objective function in an optimization process and show that ANN capture the nonlinearities of the problems. Foroud et al. [14] present a successful application of ANN combined with a genetic algorithm as optimization method for an assisted history matching study. Bruyelle and Guérillot [28] propose to compute the derivative of ANN to optimize an objective function using gradient based method. Numerous other examples of application of ANN in petroleum engineering could be cited. For example, Habiballah et al. (1995) propose to predict the vapor–liquid equilibrium ratios (K-values) for light hydrocarbon mixtures depending on component identity, mixture pressure, temperature, and convergence pressure, Guérillot [68] propose to replace the flash calculations in a compositional simulator by an ANN.

References

1. D.S. Oliver, Y. Chen, Recent progress on reservoir history matching: a review. Comput. Geosci. **15**(1), 185–221 (2011)
2. R.W. Rwechungura, M. Dadashpour, J. Kleppe, *Advanced History Matching Techniques Reviewed* (Society of Petroleum Engineers, 2011).https://doi.org/10.2118/142497-MS
3. C. Ravenne, A. Galli, H. Beucher, R. Eschard, D. Guérillot, Outcrop studies and geostatistical modelling of a middle Jurassic Brent analogue. in *Proceedings of the European Oil and Gas Conference, A Multidisciplinary Approach in Exploration and Production R&D* (London, Graham and Trotman, 1991), pp. 497–520
4. P. Corvi, K. Heffer, P. King, S. Tyson, G. Verly, C. Ehlig-Economides, I. Le Nir, S. Ronen, P. Shultz, P. Corbett, J. Lewis, G. Pickup, P. Ringrose, D. Guérillot, L. Montadert, C. Ravenne, H. Haldorsen, T. Hewett, Reservoir characterization using expert knowledge, data and statistics. Oilfield Rev. **4**, 25–31 (1992)
5. G. Chavent, M. Dupuy, P. Lemmonier, in *History Matching by Use of Optimal Theory* (Society of Petroleum Engineers, 1975). https://doi.org/10.2118/4627-PA
6. D. Guérillot, L. Pianelo, Simultaneous matching of production data and seismic data for reducing uncertainty in production forecasts. in *SPE European Petroleum Conference* (Society of Petroleum Engineers, 2000). https://doi.org/10.2118/65131-MS
7. R. Chassagne, D. Obidegwu, J. Dambrine, C. MacBeth, Binary 4D seismic history matching, a metric study. Comput. Geosci. **96**, 159–172 (2016)
8. D. Rahon, G. Blanc, D. Guérillot, Gradients method constrained by geological bodies for history matching. in *ECMOR V-5th European Conference on the Mathematics of Oil Recovery* (1996)
9. G. Blanc, D. Guérillot, D. Rahon, F. Roggero, *Building Geostatistical Models Constrained by Dynamic Data—A Posteriori Constraints* (Society of Petroleum Engineers, 1996). https://doi.org/10.2118/35478-MS
10. F. Anterion, R. Eymard, B. Karcher, *Use of parameter gradients for reservoir history matching* (Society of Petroleum Engineers, In SPE Symposium on Reservoir Simulation, 1989)
11. C. Amudo, T. Graf, R.R. Dandekar, J.M. Randle, The pains and gains of experimental design and response surface applications in reservoir simulation studies. in *SPE Reservoir Simulation Symposium* (Society of Petroleum Engineers, 2009)
12. A.S. Cullick, W.D. Johnson, G. Shi, Improved and more rapid history matching with a nonlinear proxy and global optimization. in *SPE Annual Technical Conference and Exhibition* (Society of Petroleum Engineers, 2006)
13. L.A.N. Costa, C. Maschio, D.J. Schiozer, Study of the influence of training data set in artificial neural network applied to the history matching process. in *Rio Oil and Gas Expo and Conference* (2010)
14. T. Foroud, A. Seifi, B. AminShahidi, Assisted history matching using artificial neural network based global optimization method–Applications to Brugge field and a fractured Iranian reservoir. J. Petrol. Sci. Eng. **123**, 46–61 (2014)
15. D.R. Guérillot, J. Bruyelle, *Uncertainty Assessment in Production Forecast with an Optimal Artificial Neural Network* (Society of Petroleum Engineers, 2017). https://doi.org/10.2118/183921-MS
16. E. Manceau, M. Mezghani, I. Zabalza-Mezghani, F. Roggero, Combination of experimental design and joint modeling methods for quantifying the risk associated with deterministic and stochastic uncertainties-An integrated test study. in *SPE Annual Technical Conference and Exhibition* (Society of Petroleum Engineers, 2001)
17. M. Sayyafzadeh, Reducing the computation time of well placement optimization problems using self-adaptive metamodeling. J. Petrol. Sci. Eng. **151**, 143–158 (2017)
18. J. Bruyelle, D. Guérillot, Optimization of waterflooding strategy using artificial neural networks. *SPE Reservoir Characterisation and Simulation Conference and Exhibition (RCSC 2019)* (2019)

19. J. Bruyelle, D. Guérillot, Well placement optimization with an artificial intelligence method applied to brugge field. *Gas & Oil Technology Showcase and Conference (GOTECH 2019)* (2019)
20. J. Bruyelle, D. Guérillot, Proxy model based on artificial intelligence technique for history matching-application to Brugge field. in *SPE Gas and Oil Technology Showcase and Conference.* (OnePetro, 2019). https://doi.org/10.2118/198635-MS
21. G. Cybenko, *Continuous valued neural networks with two hidden layers are sufficient.* Technical Report, Department of Computer Science, Tufts University (1988)
22. G. Cybenko, Approximation by superpositions of sigmoidal functions. in *Mathematics of Control, Signals, and Systems* (1989)
23. K. Hornik, M. Stinchcombe, H. White, Multilayer feed forward networks are universal approximators. Neural Netw. **2**, 359366 (1989)
24. K. Hornik, M. Stinchcombe, H. White, Universal approximation of an unknown mapping and its derivatives using multilayer feedforward networks. Neural Netw. **3**, 551560 (1990)
25. K. Hornik, Approximation capabilities of multilayer feedforward networks. Neural Netw. **4**, 251–257 (1991)
26. D.R. Guérillot, *EOR Screening with an Expert System* (Society of Petroleum Engineers, 1988). https://doi.org/10.2118/17791-MS
27. P.C. Silva, C. Maschio, D.J. Schiozer, Use of neuro-simulation techniques as proxies to reservoir simulator: application in production history matching. J. Petrol. Sci. Eng. **57**(3–4), 273–280 (2007)
28. J. Bruyelle, D. Guérillot, Neural networks and their derivatives for history matching and reservoir optimization problems. Comput. Geosci. **18**(3–4), 549–561 (2014)
29. D.R. Guérillot, J. Bruyelle, History matching methodology using an optimal neural network proxy and a global optimization method. in *Third EAGE Integrated Reservoir Modelling Conference* (2016). https://doi.org/10.3997/2214-4609.201602403
30. D. Guérillot, *Method and system for dynamically modeling a multiphase fluid flow*—European Patent EP2791712B1-US Patent App. 14/365,053 (2014)
31. D. Guérillot, J. Bruyelle, Geochemical equilibrium determination using an artificial neural network in compositional reservoir flow simulation. in *ECMOR XVI-16th European Conference on the Mathematics of Oil Recovery* (2018). https://doi.org/10.3997/2214-4609.201802232, to appear in Springer https://doi.org/10.1007/s10596-019-09861-4
32. S. Wang, N. Sobecki, D. Ding, L. Zhu, Y.S. Wu, Accelerating and stabilizing the vapor-liquid equilibrium (VLE) calculation in compositional simulation of unconventional reservoirs using deep learning-based flash calculation. Fuel **253**, 209–219 (2019)
33. H. Mata-Lima, Evaluation of the objective functions to improve production history matching performance based on fluid flow behaviour in reservoirs. J. Petrol. Sci. Eng. **78**(1), 42–53 (2011)
34. P. Sarma, W.H. Chen, *Efficient Well Placement Optimization with Gradient-based Algorithms and Adjoint Models* (Society of Petroleum Engineers, 2008). https://doi.org/10.2118/112257-MS
35. M. Zandvliet, M. Handels, G. van Essen, R. Brouwer, J.D. Jansen, Adjoint-based well-placement optimization under production constraints. SPE J. **13**(04), 392–399 (2008). https://doi.org/10.2118/105797-PA
36. Z. Bouzarkouna, D.Y. Ding, A. Auger, *Partially Separated Metamodels With Evolution Strategies for Well-Placement Optimization* (Society of Petroleum Engineers, 2013). https://doi.org/10.2118/143292-PA
37. N. Hansen, A. Ostermeier, Completely derandomized self-adaptation in evolution strategies. Evol. Comput. **9**(2), 159–195 (2001)
38. O.J. Isebor, L.J. Durlofsky, D.E. Ciaurri, A derivative-free methodology with local and global search for the constrained joint optimization of well locations and controls. Comput. Geosci. **18**(3–4), 463–482 (2014)
39. A.A. Awotunde, N. Sibaweihi, Consideration of voidage-replacement ratio in well-placement optimization. SPE Econ. Managem. **6**(01), 40–54 (2014). https://doi.org/10.2118/163354-PA

40. F. Forouzanfar, W.E. Poquioma, A.C. Reynolds, Simultaneous and sequential estimation of optimal placement and controls of wells with a covariance matrix adaptation algorithm. SPE J. **21**(02), 501–521 (2016). https://doi.org/10.2118/173256-PA
41. Z. Alrashdi, M. Sayyafzadeh, (μ+ λ) Evolution strategy algorithm in well placement, trajectory, control and joint optimization. J. Petrol. Sci. Eng. **177**, 1042–1058 (2019)
42. L. Montadert, La sédimentologie et l'étude détaillée des hétérogénéités d'un réservoir: application au gisement d'Hassi-Messaoud. Rev. Inst. Franc. Pétrol. Paris **17**, 241–257 (1963)
43. J. Groult, L.H. Reiss, L. Montadert, Reservoir inhomogeneities deduced from outcrop observations and production logging. J. Petrol. Technol. **18**(07), 883–891 (1966)
44. L. Tomutsa, S.R. Jackson, M. Szpakiewicz, T. Palmer, Geostatistical characterization and comparison of outcrop and subsurface facies: Shannon shelf sand ridges. in SPE California Regional Meeting (Society of Petroleum Engineers, 1986)
45. C. Ravenne, R. Eschard, A. Galli, Y. Mathieu, L. Montadert, J.L. Rudkiewicz, Heterogeneities and geometry of sedimentary bodies in a fluvio-deltaic reservoir. SPE Form. Eval. **4**(02), 239–246 (1989)
46. B. Güyagüler, R.N. Horne, *Uncertainty Assessment of Well-Placement Optimization* (Society of Petroleum Engineers, 2004). https://doi.org/10.2118/87663-PA
47. H. Wang, D. Echeverría-Ciaurri, L. Durlofsky, A. Cominelli, Optimal well placement under uncertainty using a retrospective optimization framework. SPE J. **17**(01), 112–121 (2012)
48. H. Markowitz, Portfolio selection. The J. Finan. **7**(1), 77–91 (1952)
49. B. Couet, R. Burridge, D. Wilkinson, U.S. Patent No. 6,775,578 (Washington, DC, U.S. Patent and Trademark Office, 2004)
50. A. Capolei, E. Suwartadi, B. Foss, J.B. Jørgensen, A mean–variance objective for robust production optimization in uncertain geological scenarios. J. Petrol. Sci. Eng. **125**, 23–37 (2015)
51. W.F. Sharpe, The sharpe ratio. J. Portf. Manag. **21**(1), 49–58 (1994)
52. H. Asheim, Maximization of water sweep efficiency by controlling production and injection rates. in *European Petroleum Conference* (Society of Petroleum Engineers, 1988)
53. D.R. Brouwer, J.D. Jansen, Dynamic optimization of waterflooding with smart wells using optimal control theory. SPE J. **9**(04), 391–402 (2004)
54. A. Alhuthali, A. Oyerinde, A. Datta-Gupta, Optimal waterflood management using rate control. SPE Reservoir Eval. Eng. **10**(05), 539–551 (2007)
55. A.C. Bittencourt, R.N. Horne, *Reservoir development and design optimization.* in *SPE Annual Technical Conference and Exhibition* (Society of Petroleum Engineers, 1997)
56. M. Asadollahi, G. Naevdal, *Waterflooding Optimization Using Gradient Based Methods* (Society of Petroleum Engineers, 2009). https://doi.org/10.2118/125331-MS
57. M.A.S. Pinto, C.E. Barreto, D.J. Schiozer, *Optimization of Proactive Control Valves of Producer and Injector Smart Wells under Economic Uncertainty* (Society of Petroleum Engineers, 2012). https://doi.org/10.2118/154511-MS
58. G. Van Essen, M. Zandvliet, P. Van den Hof, O. Bosgra, J.D. Jansen, *Robust Waterflooding Optimization of Multiple Geological Scenarios* (Society of Petroleum Engineers, 2009). https://doi.org/10.2118/102913-PA
59. A.H.H. Alhuthali, A. Datta-Gupta, B.B.W. Yuen, J.P. Fontanilla, Field applications of water-flood optimization via optimal rate control with smart wells. in *SPE Reservoir Simulation Symposium* (Society of Petroleum Engineers, 2009)
60. Y. Chen, D.S. Oliver, D. Zhang, Efficient ensemble-based closed-loop production optimization. SPE J. **14**(04), 634–645 (2009)
61. E. Yasari, M.R. Pishvaie, F. Khorasheh, K. Salahshoor, R. Kharrat, Application of multicriterion robust optimization in water-flooding of oil reservoir. J. Petrol. Sci. Eng. **109**, 1–11 (2013)
62. S. Hørsholt, W. Nick, J.B. Jørgensen, Oil production optimization of black-oil models by integration of Matlab and eclipse E300. IFAC-PapersOnLine **51**(8), 88–93 (2018). https://doi.org/10.1016/j.ifacol.2018.06.360

63. J. Fu, X.H. Wen, An assessment of model-based multiobjective optimization for efficient management of subsurface flow. in *SPE Western Regional Meeting* (Society of Petroleum Engineers, 2018)
64. Z. Bouzarkouna, D.Y. Ding, A. Auger, Well placement optimization with the covariance matrix adaptation evolution strategy and meta-models. Comput. Geosci. **16**(1), 75–92 (2012)
65. V.M. Johnson, L.L. Rogers, *Using artifical neutral networks and the genetic algorithm to optimize well-field design: Phase I* (No. UCRL-ID-132280) (Lawrence Livermore National Laboratory, Livermore, CA, 1998)
66. A.A. Emerick, E. Silva, B. Messer, L.F. Almeida, D. Szwarcman, M.A.C. Pacheco, M.M.B.R. Vellasco, Well placement optimization using a genetic algorithm with nonlinear constraints. Proceedings of the SPE Reservoir Simulation Symposium, The Woodlands. *Society of Petroleum Engineers*, pp. 1–20 (2009). https://doi.org/10.2118/118808-MS
67. F. Forouzanfar, A.C. Reynolds, Joint optimization of number of wells, well locations and controls using a gradient-based algorithm. Chemical Engineering Research and Design, **92**(7), 1315–1328 (2014)
68. D. Guerillot, Method and systems for reservoir modeling, evaluation and simulation. WO Patent Application WO 2012/121769 A2 (2012).

Chapter 3
Application to These Advanced Workflows to the Brugge Field Case

3.1 Description of the Brugge Field

The Brugge field is a synthetic case study submitted by TNO to compare different methods used to solve history-matching and water flooding-optimization problems in a closed-loop workflow. The complete geological description is given in Peters et al. [1]. The properties of the Bruges field are characteristic of the Brent field in the North Sea. The field is composed of four reservoir zones resulting in varying deposit environments. The reservoir properties are summarized in Table 3.1.

The model contains 30 vertical wells (20 producers and 10 injectors). The well completions correspond to the reservoir zones. The producers are drilled through the Schelde, Waal and Mass formation, while the injectors are through the whole reservoir formations. The production data of the first 10 years are available for the history matching process. For the 20 years of production optimization, the wells produce only from the Schelde and Maas formations and inject only in the Maas, Waal, and Schie formations. The economic parameters of the NPV function are given in Table 3.2.

In Annex 1, some figures shows the facies distribution.

For the Brugge field benchmark, 104 realizations of the reservoir properties were provided by TNO to participants. In these briefs, the 104 realizations provided by TNO was not used. From the geological description and the well-log data, 100 high-resolution geological models composed of 7.7 million of active grid cells were designed with an average dimension equal to 50 m × 50 m × 0.25 m. These 100 high-resolution models have been generated following the formation description: channel sand objects in a shale background in the Schelde formation; lower shoreface with carbonate concretions in Waal formation; upper shoreface in Maas formation; and Sandy Shelf with irregular carbonate patches in Schie formation. These 100 high-resolution models have been upscaled to 100 reservoir models composed of 45.500 cells of average dimensions equal to 150 m × 150 m × 7 m Fig. 3.1.

Table 3.1 Stratigraphy description of Brugge field [1]

Formation (Reservoir zone)	Average thickness [m]	Average porosity* [%]	Average permeability* [mD]	Depositional environment
Schelde	10	20.7	1105	Fluvial
Waal	26	19	90	Lower shoreface
Maas	20	24.1	814	Upper shoreface
Schie	5	19.4	36	Sandy shelf

*Average values for sand only (i.e., net porosity and permeability)

Table 3.2 Economic parameters of NPV function [1]

NPV parameters	Value
Oil price	80 $/bbl
Water production cost	5 $/bbl
Water injection cost	5 $/bbl
Annual discount rate	10%

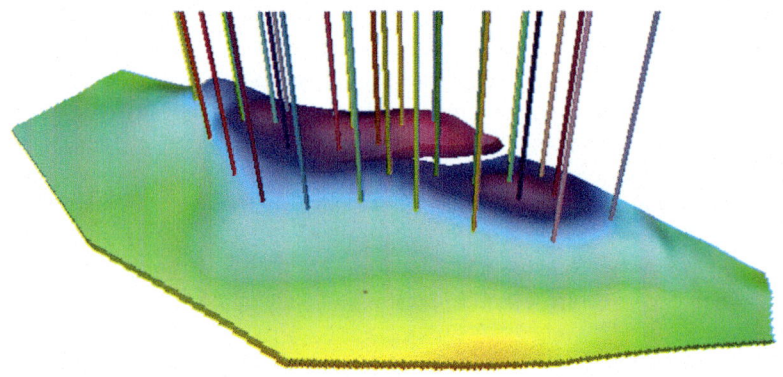

Fig. 3.1 Brugge model

3.2 History Matching

The comparison of three different proxies is presented using the Brugge field.

Fifty geological realizations have been used to represent the uncertainty of the geological model. For each lithofacies, the parameters are the porosity, permeability X, permeability Y, and permeability Z multipliers. The training dataset has been generated by running reservoir simulations with 50 the geological realizations for 761 parameters configurations defined with the Box-Behnken experimental design and 210 random samples. The validation dataset has been generated by running reservoir simulation for 50 random samples Fig. 3.2.

3.2 History Matching

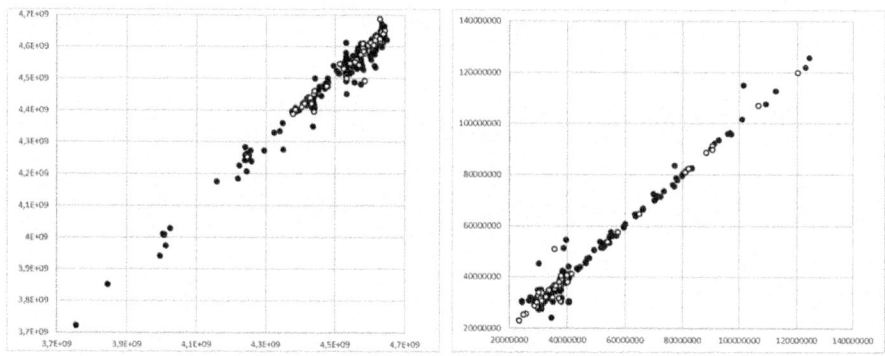

Fig. 3.2 Cross-validation plots showing the ability of the ANN to reproduce the learning data (·) and the validation data (o) for average $\mathbb{E}_\Omega[NPV]$ (left) and $\sigma_\Omega[NPV]$ (right)

The accuracy of three proxies models is compared to predict the average and standard deviation of the pressure, oil rate, and water rate at t = 27-dec-2007. The three proxies are (1) quadratic polynomial; (2) kriging; and (3) artificial neural network.

In the Annex, figures A.1–A.6 show that the ANN is more accurate than quadratic polynomial and kriging for validation dataset, i.e., data that was not used for the training phase. The root mean square mean for all proxies is given in Table 3.3.

Table 3.3 Root mean square error (Minimum error in gray)

		Average		Standard deviation	
		Learning data	Validation	Learning data	Validation
Pressure	Quadratic polynomial	2.2631	9.3322	0.5894	1.3891
	Kriging	4.9103e−13	7.5293	1.3578e−13	1.3577
	ANN	1.9485	3.0562	0.5781	0.8033
Oil rate	Quadratic polynomial	7.2751	40.4203	0.5272	2.3337
	Kriging	1.7137e−12	31.9087	1.2694e−13	2.2507
	ANN	5.4526	7.9485	0.4468	0.4820
Water rate	Quadratic polynomial	6.1149	34.3548	0.6892	3.2100
	Kriging	1.4219e−12	27.1954	1.7047e−13	3.0851
	ANN	4.4017	5.1793	0.5914	0.7623

3.3 Well Placement Optimization

The methodology is applied to identify the optimal location of these injector wells. The injection rate and the date of drilling of these wells are assumed to the same that the initial model. A comparison of results is made between (1) the production without optimization, i.e., the reference case, and (2) the production obtained with the proposed method. The proposed exercise aims to highlight that the mean-standard deviation formulation, coupled with an accurate proxy model, can provide an overview of the risk-expected return.

To generate the neural network learning database, a Box-Behnken experiment plan has been used (761 configurations) and randomly generated configurations in the parameter space (250 configurations). Fifty randomly generated configurations were used to build the validation database. For each configuration of the placement of wells, a flow simulation is performed for the 50 reservoir models assumed to represent the uncertainty associated with the geological parameters. The output data are the average of the NPV and the standard deviation of the NPV of the 50 simulations.

Figure 3.3 shows the ability of the ANN to reproduce the learning data and to predict the validation data for the average and the standard deviation of the NPV for an ensemble of geological models.

Optimization of the placement of the wells is done, for ten values of α, using the outputs of the ANN to compute the value of the objective function. Figure 3.4 shows the efficient frontier for the mean-standard deviation optimizations and associated Sharpe ratio for ten different value of alpha α. The optimized solutions for $\alpha = 0.1$ et $\alpha = 0.2$ have a Sharpe ratio between 0 and 1, i.e., a risk too high compared to the expected return. For $\alpha \geq 0.3$, the optimized solutions have a Sharpe ratio higher than one, i.e., a solution with a high expected benefit for a low taken risk. The higher Sharpe ratio is for α equals 0.3. The NPV of the optimized solution has been increased by 2.88% compared to the base case.

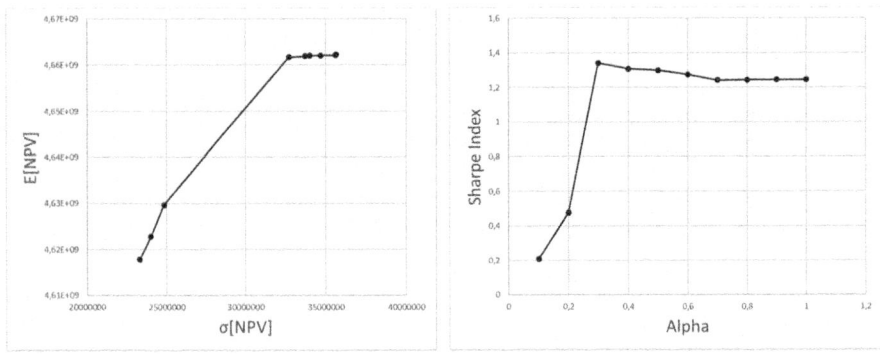

Fig. 3.3 Efficient frontier for the mean-standard deviation optimizations (left) and associated Sharpe ratio (right) for different value of *α*

3.4 Water Flooding   31

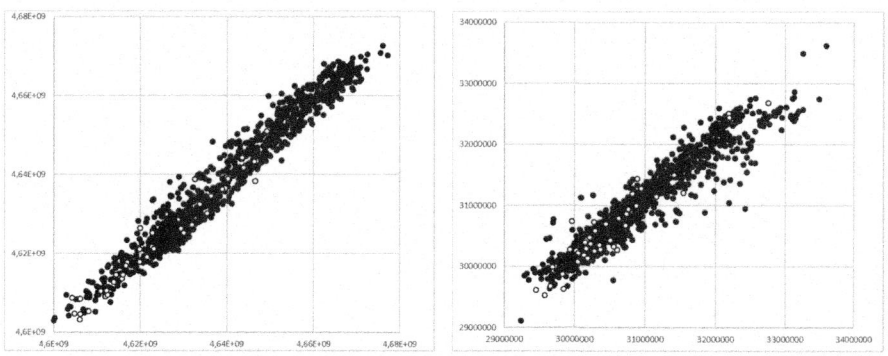

Fig. 3.4 Cross-validation plots showing the capacity of the ANN to reproduce the learning data (·) and the validation data (○) for average $\mathbb{E}_\Omega[NPV]$ (left) and $\sigma_\Omega[NPV]$ (right)

3.4 Water Flooding

The methodology of waterflooding optimization has been applied for five wells to the period 10–20 years. The control parameter for the three producers (P11, P12, and P13) and the two injectors (I8 and I9) are the production/injection rates updated each year. All other wells have been closed. The choice of these producer wells has to be done due to the high watercut value after ten years of production, and the two injector wells are the closest to these producers. The 50 realizations that best fit the history matching data has been selected to represent the uncertainty of the geological model.

The training dataset has been generated by running reservoir simulations with the 50 geological realizations for 1200 random configurations of production/injection and a part of Box-Behnken experimental design (850 configurations). The validation dataset has been generated by running reservoir simulation for 50 random configurations of production/injection. The ANN is defined with seven neurons in the hidden layer.

The cross-validation plots (Fig. 3.3) show the capacity of the ANN to reproduce the average and standard deviation of the NPV for both the learning data and the validation data. However, we note that the prediction of the standard deviation is not as accurate as wanted. In order to improve the predictive value of the standard deviation with the ANN, it would be necessary to increase the number of training data in order to increase the number of neurons in the hidden layer and thus increase the number of degrees of freedom of the proxy.

After the learning phase, an optimization process is run with the CMA-ES as the optimizer and the objective function defined with the outputs of the ANN, for ten values of α. Figure 3.4 shows the efficient frontier for the mean-standard deviation optimizations, i.e., the 'risk-expected return' optimized solutions for different value of α. The more α increases, the more the expected gain and associated risk increases. The higher Sharpe ratio is obtained with $\alpha = 0.3$. As we consider the risk-free

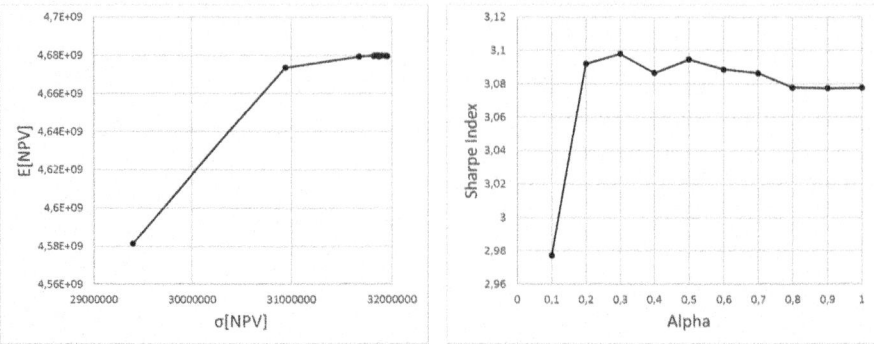

Fig. 3.5 Efficient frontier for the mean-standard deviation optimizations (left) and associated Sharpe ratio (right) for different value of α

configuration as the one that minimizes the standard deviation (i.e., the optimized configuration for $\alpha = 0$), the Sharpe ratio for α equal zero is null Fig. 3.5.

Acknowledgements The author expresses his deepest gratitude to his co-author, Jérémie Bruyelle, now with CEA Cadarache, France, whose contributions were instrumental to the success of the work presented in this book. His deep expertise in artificial intelligence techniques and stochastic optimization formed the backbone of the methodologies developed and applied throughout this project. Without his talent and dedication, this body of work would not have been possible.

This project has been generously supported by Terra 3E SAS (http://www.Terra3E.com), a company founded by the author in 2009, and by Texas A&M University at Qatar. The author gratefully acknowledges Schlumberger (SLB) for their in-kind donation of Petrel® software and the Eclipse 100® simulator, and Terra 3E for providing access to its proprietary OPUS TERRA™ software, which includes modules for artificial neural networks and optimization.

The author also acknowledges the Society of Petroleum Engineers (SPE) for granting permission to reuse material from the following previously published papers: SPE-183921-MS, SPE-196643-MS, SPE-198635-MS, and SPE-198656-MS. These papers, co-authored with Jérémie Bruyelle and others, served as the foundation for several chapters in this book. The author also thanks SPE for its continued support in disseminating technical knowledge within the oil and gas industry.

Reference

1. L. Peters, R.J. Arts, G.K. Brouwer, C.R. Geel, S. Cullick, R.J. Lorentzen, Y. Chen, K.N.B. Dunlop, F.C. Vossepoel, R. Xu, P. Sarma, A.H. Alhutali, A.C. Reynolds, results of the brugge benchmark study for flooding optimization and history matching. SPE Reserv. Evalu. Eng. **13**(3), 391–405 (2010)

Thesaurus

To help the reader who may not be familiar with all concepts used in these briefs, we have added a kind of short dictionary named thesaurus to detail these concepts. We have extensivily used ChatGPT for this part after reviewing its responses...(of course!):

1. Articial Neural Network (ANN)
2. Digital twins and proxy models
3. Geostatistics
4. History matching
5. Proxy model
6. Water-flooding

Artificial Neural Network (ANN)

An Artificial Neural Network (ANN) is a computational model inspired by the structure and function of the human brain, consisting of interconnected nodes that can learn and make predictions based on data. ANNs are a subset of machine learning and are widely used in various applications, such as pattern recognition, image and speech recognition, natural language processing, and financial analysis.

The basic building block of an ANN is a node, or neuron, which receives input signals from other neurons or external sources, applies a nonlinear transformation to them, and produces an output signal that can be passed to other neurons. The neurons are organized into layers, with each layer performing a specific type of computation. The input layer receives the raw data, the output layer produces the final prediction or decision, and the hidden layers perform intermediate computations to extract features and patterns from the input data.

The training process of an ANN involves adjusting the strengths of the connections between the neurons, or weights, to minimize the difference between the predicted output and the actual output for a given set of inputs. This process typically involves backpropagation, where the error is propagated backwards through the network, and the weights are updated using an optimization algorithm such as stochastic gradient descent.

Some key characteristics of ANNs include:

Nonlinearity: ANNs can capture complex nonlinear relationships between input and output variables, making them well-suited for modeling nonlinear phenomena and performing tasks such as classification and regression.

Generalization: ANNs can learn to generalize from a set of training examples to make predictions on unseen data. This is achieved by avoiding overfitting, or memorizing the training data, and instead learning to identify the underlying patterns and relationships.

Parallelism: ANNs can perform computations in parallel, with many neurons operating simultaneously. This allows for faster processing and scalability to large datasets.

There are various types of ANNs, each with its own architecture and application. Some common types include:

Feedforward neural networks: These are the simplest type of ANN, consisting of a series of layers that process the input data in a forward direction without loops or feedback.

Recurrent neural networks: These networks have connections between neurons that form loops, allowing them to process sequential data such as time series or natural language.

Convolutional neural networks: These networks are designed to process spatial data such as images, using specialized layers that can detect local patterns and features.

ANNs have become increasingly popular in recent years, due to their ability to learn from large and complex datasets and perform a wide range of tasks with high accuracy. However, they also require careful tuning and training to achieve optimal performance, and their internal workings can be difficult to interpret.

Digital Twins and Proxy Models

Digital twins are virtual replicas of physical objects, systems, or processes that are used to simulate, analyze, and optimize their real-world counterparts.

Our proxy model could be seen as a digital twins which help making decision on geoscience applications.

Digital twins continuously receive data from their physical counterparts, allowing them to reflect current conditions, states, and performance. Here, the « physical part is not the real world but the results of the reservoir simulator.

Digital twins and proxy models are both used to simulate and analyze systems, but they serve different purposes and are built with different levels of complexity and accuracy. Here's a breakdown of their key differences:

Digital twins are comprehensive, real-time virtual replicas of physical systems or processes. They are designed to mirror the exact state of the physical counterpart and are used for real-time monitoring, diagnostics, optimization, and prediction. They are used in complex systems where real-time data integration and interaction are critical, such as in manufacturing, smart cities, healthcare, and energy management. They allow for dynamic and continuous feedback and control.

Proxy models, also known as surrogate models, are simplified representations of complex systems. They are used primarily for faster simulations when the full model (e.g., a high-fidelity simulation) is too computationally expensive or time-consuming to run. Proxy models approximate the behavior of the system based on a subset of the data or simplified assumptions.

They are often used in optimization problems, uncertainty quantification, and sensitivity analysis, particularly in fields like reservoir engineering, where running full-scale simulations repeatedly would be impractical.

Geostatistics

Geostatistics is a branch of statistics that deals with the analysis and interpretation of spatial or geospatial data. It provides a set of tools and techniques for studying and modeling spatial patterns and variability in data that are distributed over geographic locations.

The inventor of these concepts is Georges Matheron, a french scientist from Paris School of Mines.

Matheron's groundbreaking work in the 1960s and 1970s laid the groundwork for geostatistics as a scientific discipline. He developed the theoretical framework for analyzing and modeling spatial data, including the concepts of spatial dependence, the variogram, and kriging. These concepts provided a systematic way to study the spatial structure of data and make predictions at unobserved locations.

Matheron's contributions extended beyond theory, as he also developed practical applications of geostatistics. He applied geostatistical methods to various fields, including mining, hydrology, and geology, and demonstrated their effectiveness in solving real-world problems. His work played a crucial role in establishing geostatistics as a valuable tool in geosciences and other disciplines dealing with spatial data.

Georges Matheron's pioneering efforts in geostatistics continue to shape the field, and his work is widely acknowledged and celebrated by researchers and practitioners in the geospatial community. His contributions have had a lasting impact on the theory, methodology, and applications of geostatistics, making him an influential figure in the development of this scientific field.

Geostatistics is widely used in various fields, including environmental sciences, geology, mining, hydrology, agriculture, and urban planning. It allows researchers and analysts to understand the spatial structure of data, make predictions or estimates at unobserved locations, and assess the uncertainty associated with those predictions.

The fundamental concept in geostatistics is spatial dependence or spatial autocorrelation. It refers to the notion that nearby locations tend to have similar values, while locations farther apart may exhibit different values. Geostatistical methods use statistical models to quantify and characterize this spatial dependence.

Some key concepts and techniques in geostatistics include:

Variogram: The variogram is a measure of spatial dependence and describes how the similarity between data points changes with distance or lag. It is used to model the spatial correlation structure of the data.

Kriging: Kriging is a spatial interpolation method used to estimate values at unobserved locations based on observed data. It incorporates both the spatial correlation structure (as described by the variogram) and the observed values to make predictions. Kriging provides optimal, unbiased estimates with minimum prediction variance.

Spatial modeling: Geostatistics allows for the modeling of spatial processes, where the spatial dependence is accounted for in the statistical model. This enables the creation of realistic representations of spatial phenomena and the ability to generate spatial simulations.

Spatial analysis: Geostatistics provides a framework for analyzing spatial patterns and relationships. It includes techniques such as cluster analysis, spatial regression, and spatial autocorrelation analysis, which help identify trends, hotspots, and spatial relationships between variables.

Geostatistical software: Several software packages, such as Geostatistical Analyst in ArcGIS, R packages (e.g., gstat, geoR), and specialized software like GSLIB and Stanford Geostatistical Modeling Software (SGeMS), provide tools for implementing geostatistical methods and analyzing spatial data.

By applying geostatistical techniques, analysts can gain valuable insights into the spatial characteristics of data, make informed decisions in spatial planning and resource management, and improve the accuracy of predictions in various geospatial applications.

Geovariances is a company that specializes in geostatistics and provides software solutions for geostatistical analysis and modeling. They offer several software packages that are widely used in the geostatistics community. Here are a few notable software offerings by Geovariances:

Isatis: Isatis is a comprehensive software package for geostatistical analysis and mapping. It provides tools for variography, kriging, simulation, and uncertainty assessment. Isatis offers a user-friendly interface, allowing users to perform a wide range of geostatistical tasks efficiently.

Minestis: Minestis is specifically designed for the mining industry. It offers advanced geostatistical methods for resource estimation, grade control, and risk analysis in mineral exploration and mining projects. The software includes features such as block modeling, conditional simulation, and geostatistical simulations.

Kartotrak: Kartotrak is a software solution for contaminated site characterization and risk assessment. It combines geostatistics with advanced geospatial analysis techniques to support decision-making in environmental remediation projects. Kartotrak offers capabilities for spatial data analysis, modeling, and visualization.

Geovariances' software packages are widely used in industries such as mining, oil and gas, environmental consulting, and geosciences research. They provide powerful tools for analyzing spatial data, characterizing uncertainty, and making informed decisions based on geostatistical analyses.

It's important to note that while Geovariances offers specialized software solutions, there are also other popular geostatistical software options available in the industry. These include open-source software like R packages (e.g., gstat, geoR) and commercial software such as Geostatistical Analyst in ArcGIS, SGeMS, and GSLIB. The choice of software depends on the specific requirements of the project, the user's familiarity with the software, and the available resources.

Petrel, developed by Schlumberger's Software Integrated Solutions, is a widely used software platform in the oil and gas industry for geologic modeling, reservoir characterization, and reservoir simulation. It provides advanced geostatistical tools and workflows that support the analysis and modeling of subsurface data.

Petrel offers a range of geostatistical functionalities that enable users to integrate various data types, including well logs, seismic data, and production data, into a unified geologic model. Some key geostatistical features of Petrel include:

Variogram analysis: Petrel allows users to analyze spatial correlation and variability in their subsurface data by computing and visualizing variograms. Variograms provide insights into the spatial continuity of reservoir properties and can be used to guide subsequent modeling workflows.

Kriging and interpolation: Petrel supports kriging and other interpolation methods to estimate reservoir properties at unobserved locations. These methods leverage the spatial correlation structure of the data to provide reliable predictions across the reservoir.

Geological modeling: Petrel's geostatistical tools can be applied to create detailed geological models. These models help capture the spatial heterogeneity of reservoir properties, which is crucial for accurate reservoir characterization and simulation.

Uncertainty assessment: Petrel facilitates uncertainty analysis by integrating geostatistical simulations with stochastic modeling techniques. Users can generate multiple realizations of the reservoir model to assess the uncertainty in predicted reservoir behavior.

Petrel is known for its user-friendly interface and its ability to handle large and complex datasets. It is widely used by geoscientists and reservoir engineers in the oil and gas industry for optimizing field development strategies, making informed decisions, and maximizing hydrocarbon recovery.

History Matching

History matching is a technique used in the field of reservoir engineering to calibrate and validate reservoir models by comparing their simulated behavior with observed historical data. It is an important step in the process of reservoir simulation and optimization.

Reservoir models are mathematical representations of underground oil or gas reservoirs, which are used to predict their behavior and estimate the production performance. These models incorporate various parameters, such as reservoir geometry, petrophysical properties, fluid behavior, and production mechanisms, among others.

However, due to uncertainties and limitations in data and knowledge, reservoir models often have discrepancies when compared to actual reservoir behavior. This is where history matching comes into play. The goal of history matching is to adjust the uncertain parameters within the reservoir model to minimize the discrepancies between simulated results and observed field data.

The history matching process involves several steps:

Data Collection: Relevant historical field data, such as production rates, pressure measurements, fluid compositions, and any other available information, is gathered.

Reservoir Simulation: The initial reservoir model is constructed using available data and knowledge. This model is then used to simulate the reservoir behavior and generate synthetic data for comparison with the historical data.

Objective Function Definition: An objective function is defined to quantify the mismatch between simulated results and observed data. It typically includes a combination of different types of data, such as production rates, pressure differentials, and water-cut percentages. The objective function represents the measure of the model's quality.

Uncertainty Analysis: The uncertain parameters within the reservoir model, such as permeability values, porosity, and initial fluid saturations, are identified. These parameters are varied within a certain range to account for their uncertainties.

Optimization: An optimization algorithm, such as a history matching algorithm or a Bayesian inference method, is employed to find the optimal values of the uncertain parameters that minimize the objective function. The optimization process involves iteratively adjusting the parameters, running reservoir simulations, and comparing the results until an acceptable match is achieved.

Model Validation: Once a satisfactory match is obtained between the simulated results and the historical data, the calibrated reservoir model is considered valid. It can then be used for future predictions and decision-making, such as optimizing production strategies, estimating reserves, and assessing field development plans.

History matching is a challenging and computationally intensive task as it involves exploring a high-dimensional parameter space and solving complex optimization problems. It requires expertise in reservoir engineering, statistical analysis, and numerical simulation techniques. However, it is an essential tool for improving the accuracy and reliability of reservoir models, leading to better reservoir management and economic decision-making in the oil and gas industry.

In history matching, it is crucial to recognize that there are often multiple sets of parameter values that can provide a reasonably good match between the simulated results and observed data. This is known as non-uniqueness.

Non-uniqueness arises due to various factors, including the complex and nonlinear nature of reservoir systems, the limited amount and quality of available data, and the presence of uncertainties in the model and measurements. Different combinations of uncertain parameters can result in similar matches to historical data, making it challenging to identify the true underlying reservoir characteristics.

Furthermore, even if a history match is achieved, it does not guarantee that the calibrated reservoir model will provide accurate forecasts of future behavior. The future performance of a reservoir is influenced by numerous dynamic factors, such as changing reservoir conditions, production strategies, and uncertainties in the geological and fluid behavior.

A good history match should be seen as a starting point rather than a definitive solution. Sensitivity analysis and uncertainty quantification techniques are often employed to assess the range of possible outcomes and evaluate the robustness of the calibrated model. Monte Carlo simulations, ensemble-based methods, or other probabilistic approaches are used to explore the uncertainty space and generate multiple forecasts that capture the range of possible reservoir behaviors.

By considering the non-uniqueness of history matching results and accounting for uncertainties in forecasts, reservoir engineers can make more informed decisions, evaluate the associated risks, and develop appropriate strategies for reservoir management and production optimization.

Proxy Model

A proxy model, also known as a surrogate model or metamodel, is a simplified mathematical or statistical model that is used to approximate the behavior of a more complex and computationally expensive model. It serves as a substitute or "proxy" for the original model when performing simulations, optimizations, or other analyses.

Proxy models are commonly used in situations where the original model is time-consuming or computationally intensive to run, making it impractical for certain tasks. By constructing a simpler proxy model based on a subset of inputs and outputs from the original model, analysts can obtain approximate results with significantly reduced computational costs.

The process of creating a proxy model typically involves the following steps:

Sample generation: A set of input values is selected or generated to run simulations on the original model. These input–output pairs are used to build the proxy model.

Model fitting: The proxy model is constructed using mathematical or statistical techniques that capture the relationships between the input variables and the corresponding outputs from the original model. Common approaches include regression analysis, neural networks, support vector machines, or Gaussian processes.

Model validation: The accuracy and reliability of the proxy model are assessed by comparing its predictions to the outputs of the original model for a separate validation dataset. This step ensures that the proxy model adequately approximates the behavior of the original model.

Once the proxy model is constructed and validated, it can be used for various purposes, including:

Prediction: The proxy model can be used to quickly estimate the outputs of the original model for new sets of inputs, without the need to run the computationally expensive model each time.

Optimization: Proxy models are often employed in optimization algorithms to find the optimal values of input variables while minimizing or maximizing a specific objective function. The proxy model enables faster iterations of the optimization process.

Sensitivity analysis: Proxy models can be used to assess the impact of input variables on the outputs of the original model. This helps in understanding the relative importance of different factors and guiding decision-making.

Proxy models offer a trade-off between accuracy and computational efficiency. While they may not capture all the nuances of the original model, they provide a cost-effective approach to obtain reasonably accurate results in a timely manner. Proxy modeling is particularly valuable in situations where real-time or near-real-time analysis is required or when performing extensive simulations or optimizations with limited computational resources.

Water Flooding

Water flooding, also known as water injection or waterflood, is a commonly used secondary recovery method in the oil and gas industry. It involves injecting water into an oil reservoir to displace and push the oil towards production wells, thereby increasing the overall recovery of oil from the reservoir.

The process of water flooding typically follows these steps:

Reservoir characterization: The reservoir is initially studied and characterized to understand its geological and fluid properties. This includes analyzing the reservoir's structure, porosity, permeability, and fluid saturation.

Water injection well installation: Injection wells are drilled and completed in strategic locations within the reservoir. These wells are specifically designed to inject water into the reservoir at high pressures.

Water injection: Water is injected into the reservoir through the injection wells. The injected water serves multiple purposes: it maintains reservoir pressure, displaces oil, and sweeps it towards production wells.

Oil production: As water is injected into the reservoir, it displaces and pushes the oil towards production wells. The oil is then produced from these wells, along with the produced water.

Monitoring and optimization: The water flooding process is continuously monitored and optimized to maximize oil recovery. This includes monitoring injection and production rates, analyzing reservoir performance, adjusting injection patterns, and evaluating the effectiveness of the water flooding technique.

Water flooding works by utilizing the mobility difference between water and oil. Water has a higher mobility or ability to flow through the reservoir than oil. When water is injected, it displaces the oil and drives it towards production wells. This helps to sweep the oil from the reservoir, improving the overall recovery factor.

Water flooding can be enhanced through various techniques, such as:

Pattern flooding: Water is injected into the reservoir in a patterned manner, targeting specific areas of the reservoir to optimize oil displacement.

Chemical flooding: Chemicals, such as surfactants or polymers, can be added to the injected water to improve oil recovery by reducing interfacial tension or increasing sweep efficiency.

WAG (Water Alternating Gas) flooding: In addition to water, gas (such as carbon dioxide or nitrogen) is alternately injected into the reservoir to improve displacement and sweep efficiency.

Water flooding is a widely used and effective technique for increasing oil recovery from reservoirs. It can significantly improve the recovery factor, especially in reservoirs where primary production methods have reached their limits. However, the success of water flooding depends on the reservoir's characteristics, the injection strategy, and proper monitoring and optimization throughout the process.

Annex 1

(Figures A.1, A.2, A.3, A.4, A.5 and A.6).

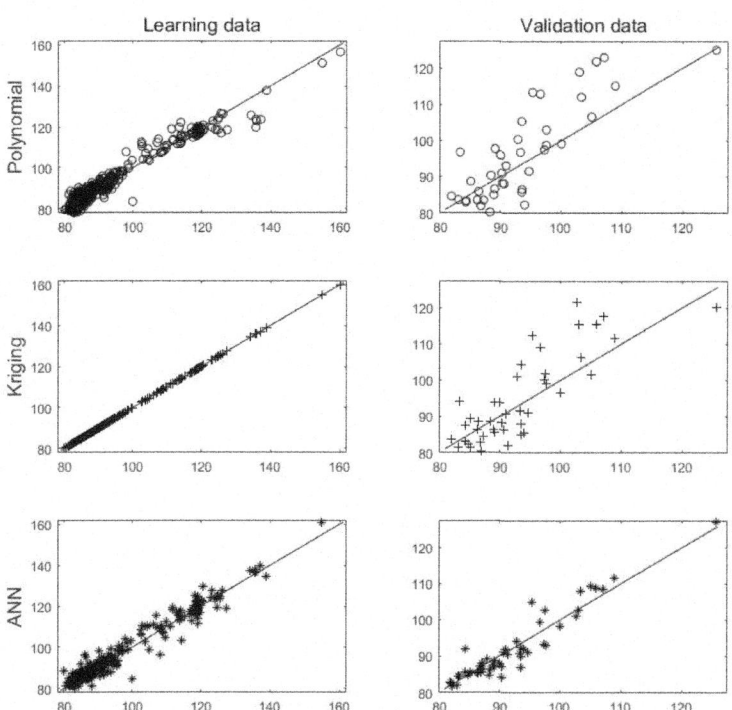

Fig. A.1 Average pressure at t = 27-Dec-2007—Cross-validation plots for quadratic polynomial (Top), kriging (Middle) and ANN (Bottom) for learning data (Left), and validation data (Right)

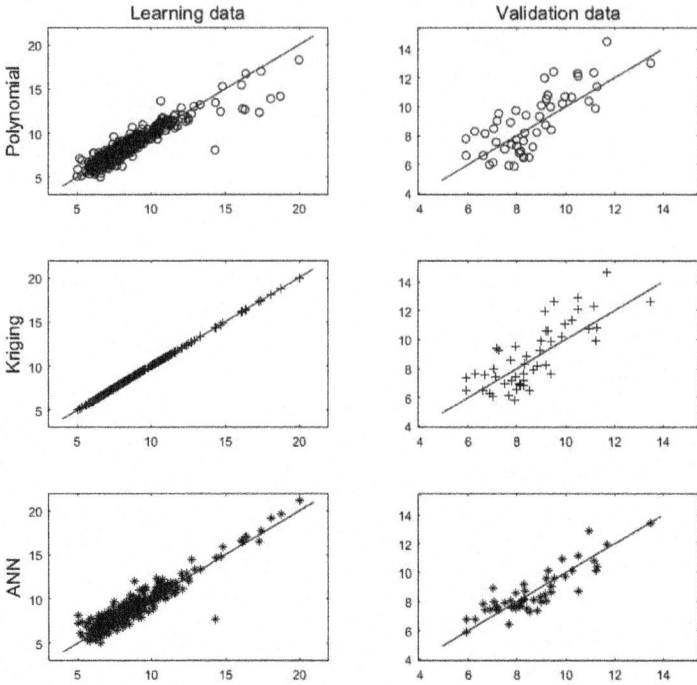

Fig. A.2 Standard deviation pressure at t = 27-Dec-2007—Cross-validation plots for quadratic polynomial (Top), kriging (Middle) and ANN (Bottom) for learning data (Left), and validation data (Right)

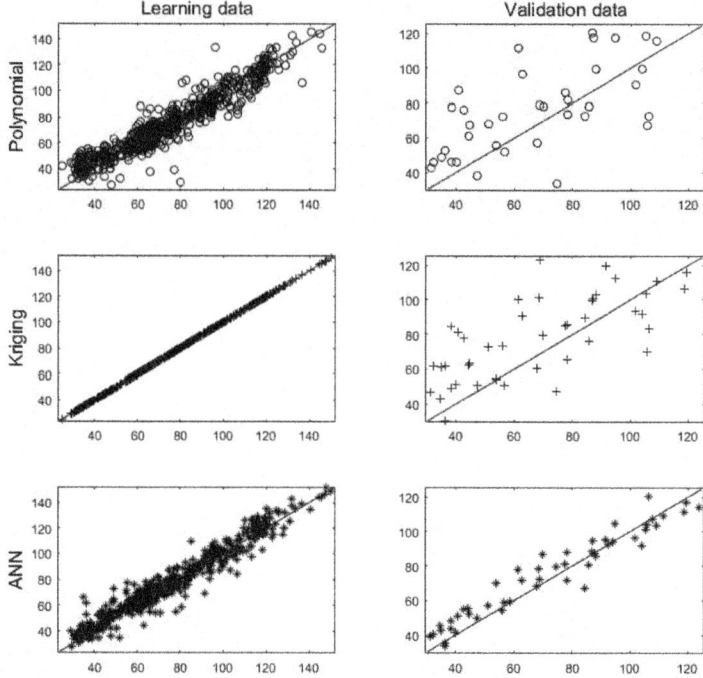

Fig. A.3 Average oil rate at t = 27-Dec-2007—Cross-validation plots for quadratic polynomial (Top), kriging (Middle) and ANN (Bottom) for learning data (Left), and validation data (Right)

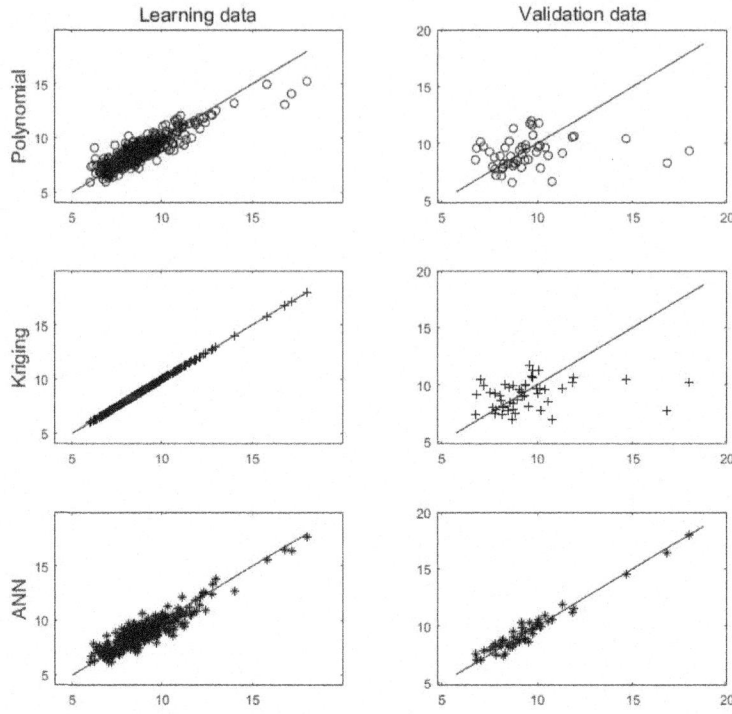

Fig. A.4 Standard deviation oil rate at t = 27-Dec-2007—Cross-validation plots for quadratic polynomial (Top), kriging (Middle) and ANN (Bottom) for learning data (Left), and validation data (Right)

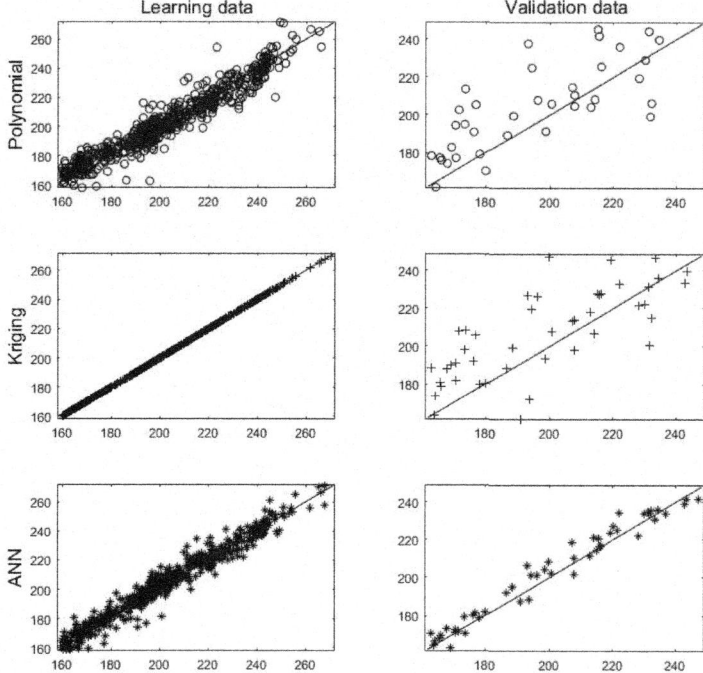

Fig. A.5 Average water rate at t = 27-Dec-2007—Cross-validation plots for quadratic polynomial (Top), kriging (Middle) and ANN (Bottom) for learning data (Left), and validation data (Right)

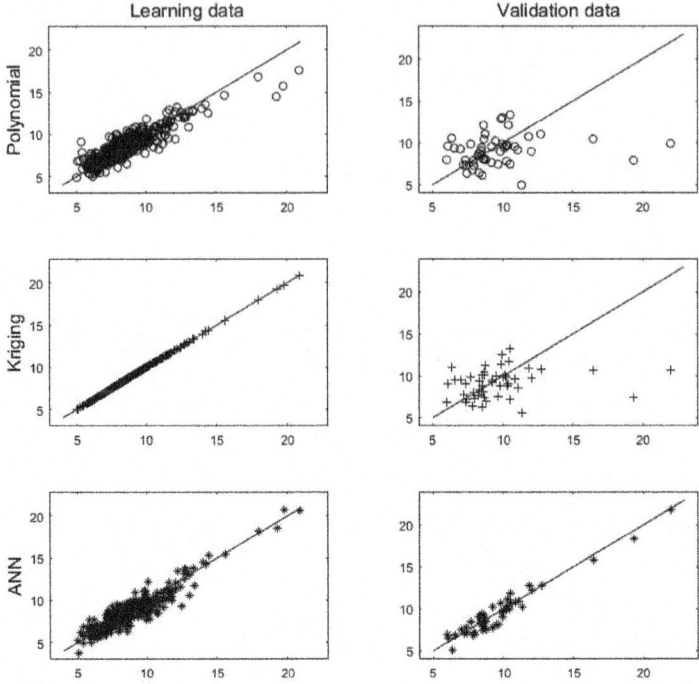

Fig. A.6 Standard deviation water rate at t = 27-Dec-2007—Cross-validation plots for quadratic polynomial (Top), kriging (Middle) and ANN (Bottom) for learning data (Left), and validation data (Right)

Annex 1

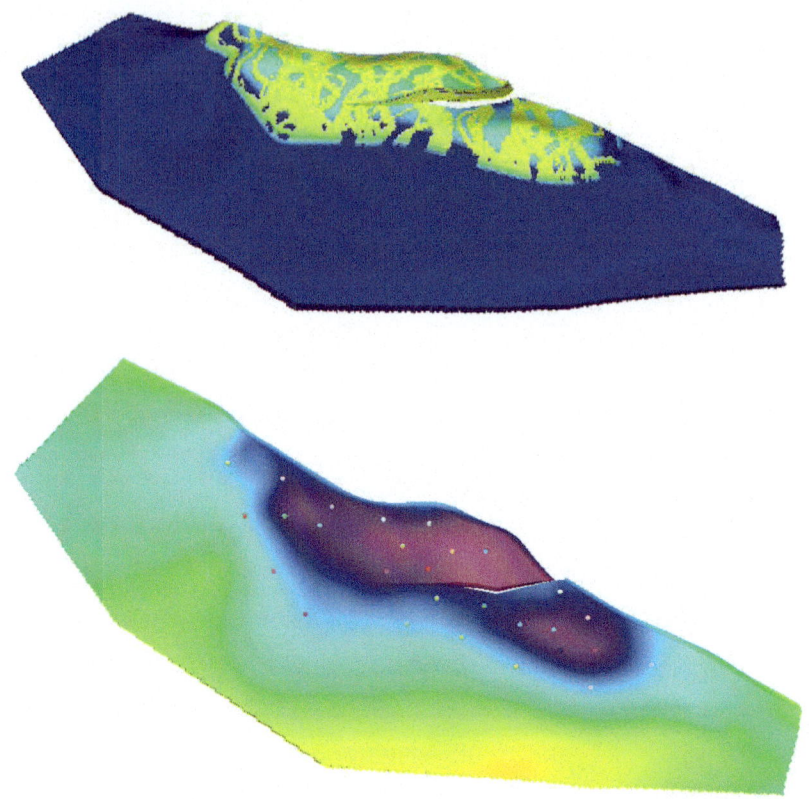

https://pubs.acs.org/doi/full/10.1021/acs.energyfuels.3c04847?casa_token=hdUgRknD_YIAAAAA%3AgS4El9sCz9ZPg-4uF1nwfRywJ64BnQR16ZzUZKzz4BwYTolAeV5L1ksibVym_qigREUKLcH60haWCWbE.

Bibliography

1. S. Amini, S. Mohaghegh, Application of machine learning and artificial intelligence in proxy modeling for fluid flow in porous media. Fluids **4**(3), 126 (2019)

The manufacturer's authorised representative in the EU is Springer Nature Customer Service Centre GmbH, Europaplatz 3, 69115 Heidelberg, Germany. If you have any concerns regarding our products, please contact ProductSafety@springernature.com

Printed and bound by CPI Group (UK) Ltd, Croydon, CR0 4YY
27/03/2026
02079806-0001